DISCARDED

The Nationalization of Venezuelan Oil

James F. Petras
Morris Morley
Steven Smith

The Praeger Special Studies program—utilizing the most modern and efficient book production techniques and a selective worldwide distribution network—makes available to the academic, government, and business communities significant, timely research in U.S. and international economic, social, and political development.

The Nationalization of Venezuelan Oil

HD
9574
V42
P46

PRAEGER SPECIAL STUDIES IN INTERNATIONAL ECONOMICS AND DEVELOPMENT

Praeger Publishers New York London

Library of Congress Cataloging in Publication Data

Petras, James F   1937-
    The nationalization of Venezuelan oil.

    (Praeger special studies in international economics
and development)
    Includes index.
    1. Petroleum industry and trade--Government ownership
--Venezuela. I. Morley, Morris, joint author.
II. Smith, Steven, joint author. III. Title.
HD9574.V42P46   338.2'7'2820987   77-7822
ISBN 0-03-022656-2

PRAEGER SPECIAL STUDIES
200 Park Avenue, New York, N.Y., 10017, U.S.A.

Published in the United States of America in 1977
by Praeger Publishers,
A Division of Holt, Rinehart and Winston, CBS, Inc.

789   038   987654321

© 1977 by Praeger Publishers

All rights reserved

Printed in the United States of America

In memory of
Orlando Letelier,
friend, fighter and scholar.

ACKNOWLEDGMENTS

The research and writing of this study benefited from the assistance of a great number of people, foremost among whom was Orlando Letelier, who encouraged us and was very helpful setting up interviews and later in providing us with critical comments on the manuscript. We benefited through discussion and criticism from many other individuals, including Pedro P. Azpurua, Allan R. Brewer-Carias, Michel Chossudovsky, Julio Esteves, Marcos Negron, Jose Rafael Tenorio, Freddy Munoz, Francisco Mieres, Americo Martin, Armando Cordova, Alfredo Chacon, Hector Silva Michelena, Luis Lander and Fernando Travieso. We are especially indebted to the works of Domingo Alberto Rangel. The authors are ultimately responsible for the integration and perspectives presented.

CONTENTS

	Page
ACKNOWLEDGMENTS	vi
LIST OF TABLES	ix
LIST OF ABBREVIATIONS	x
CHRONOLOGY OF VENEZUELAN GOVERNMENTS SINCE 1870	xi
CHRONOLOGY OF VENEZUELAN GOVERNMENT OIL POLICIES	xii
INTRODUCTION	xiii

Chapter

1 CAPITALIST DEVELOPMENT IN HISTORIC PERSPECTIVE ... 1

    Introduction ... 1
    Legacy of the Imperialism of Free Trade ... 2
    Military Mercantilism ... 5
    The Failure of National Populism ... 8
    The Alienated State ... 15
    The Consolidation of State Capitalism ... 26
    Conclusion ... 42
    Notes ... 44

2 NATIONALIZATION AND CAPITALIST DEVELOPMENT ... 49

    Introduction ... 49
    Aspects of Nationalization in the Third World ... 50
    State Capitalism ... 58
    Nationalization and the International Conjuncture ... 77
    Impediments to Growth and Equity: The Primacy of the Political ... 77
    Conclusion ... 90
    Notes ... 93

3 NATIONALIZATION AND U.S. POLICY ... 95

    Introduction ... 95
    U.S. Business Community ... 104

Oil Companies	114
U.S. State Department	123
U.S. Department of the Treasury	130
U.S. Department of Defense	134
U.S. Congress	138
International Banks	141
Notes	145

4 SUMMARY 151

The Transition to Industrial Capitalism	151
Class and State	153
The Nationalization	155
The United States, Venezuela, and the International Politics of Oil	158
Notes	159

APPENDIX: RECENT TRENDS 161

Nationalists with Money	161
International Debt Peonage	162
The Limits of Trickle-Down Growth	163
Behind the Nationalist Facade	165
Notes	168

INDEX 171

ABOUT THE AUTHORS 175

LIST OF TABLES

Table		Page
1	Typology of State Capitalist Nationalization	51
2	U.S. Foreign Trade with Venezuela, Selected Commodities	109
3	U.S. Direct Investment in Selected Latin American Countries, 1966–74, Total Manufacturing	110
4	U.S. Direct Investment in Venezuela, 1966–74	111
5	Summary Trends in Venezuelan Prices	164

LIST OF ABBREVIATIONS

AD	Democratic Action Party
BCV	Central Bank of Venezuela
BIV	Venezuelan Industrial Bank
CADAFE	Company for the Administration & Supply of Electricity
CARICOM	Caribbean Community
CENDES	Center for Development Studies
CNT	National Confederation of Workers
CODESA	Committee of Autonomous Unions
COPEI	Christian Democratic Party
CORDIPLAN	Central Office of Coordination and Planning
CORPOINDUSTRIA	Corporation for the Promotion of Small and Medium Industry
CTV	Confederation of Venezuelan Workers
CUTV	Unified Confederation of Venezuelan Workers
CVF	Venezuelan Development Corporation
CVG	Venezuelan Guyana Corporation
CVP	Venezuelan Petroleum Corporation
DIVIDENDO	Voluntary Dividends for the Community
EDUPLAN	Office of Integrated Educational Planning
FCV	Federation of Venezuelan Peasants
Fedecámaras	Federation of Chambers of Commerce and Production
Fedepetrol	Federation of Petroleum Workers
FEI	Independent Electoral Front
FIV	Venezuelan Investment Fund
IAN	National Agrarian Institute
IVP	Venezuelan Petrochemical Institute
LAFTA	Latin American Free Trade Association
MAS	Movement to Socialism
MEP	People's Electoral Movement
MIR	Movement of the Revolutionary Left
OPEC	Organization of Petroleum Exporting Countries
PETROVEN	Petroleum Company of Venezuela
PRIN	Revolutionary Party of Nationalist Integration
SELA	Latin American Economic System
URD	Democratic Republican Union
VIPOSA	Popular Homes Corporation

CHRONOLOGY OF VENEZUELAN GOVERNMENTS
SINCE 1870

António Guzmán Blanco	1870–88
Joaquín Crespo	1889–98
Cipriano Castro	1899–1907
Juan Vicente Gómez	1908–35
Eleazar López Contreras	1936–41
Isaías Medina Angarita	1942–45
Provisional Government	1945–46
Rómulo Gallegos	1947–48
Carlos Delgado Chalbaud and Marcos Pérez Jiménez	1948–50
Marcos Pérez Jiménez	1950–57
Provisional Government	1958
Rómulo Betancourt	1959–63
Raúl Leoni	1964–68
Rafael Caldera	1969–73
Carlos Andrés Pérez	1974–

CHRONOLOGY OF VENEZUELAN GOVERNMENT OIL POLICIES

Concessions System
1. Concept of private oil concessions established in law (1920).
2. Hydrocarbons Law (1943) created uniform tax treatment for all concessions and increased the state's participation in oil companies' profits to 50 percent.
3. Sale of further concessions ended (1947).
4. Previous law revoked; new concessions sold (1956–57).

Service Contract System
1. OPEC formed as international oil cartel (1960).
2. Venezuelan Petroleum Corporation established (1960) to administer the service contracts (direct contracting out to foreign companies or combining with such corporations to form mixed companies), to participate directly in oil-related activities (exploration, extraction, refining, marketing, transport, and so on), and to subsidize private firms in the oil industry.
3. Subsoil rights retained and the further sale of concessions prohibited.
4. Level of the state's participation in oil companies' profits raised to 78 percent (1971).
5. Oil Reversion Law (1971) sets date for the end of all current concessions and expropriates those concessions that are not currently being exploited.
6. Natural gas nationalized (1971).
7. Petroleum nationalized (1976).

INTRODUCTION

OUTLINE

This study focuses on one of the most crucial issues of our time, the exploitation and control of one of the most strategic resources of the industrialized countries—oil.

For many decades ownership by Western countries facilitated plentiful supplies of petroleum at cheap prices. As a result, industrial empires grew while the oil-producing countries remained poor (except for tiny ruling elites) and at the periphery of world politics. Today, the rise of nationalism in the Third World has found one of its principal spokesmen in the oil-producing countries. Through association in a producer cartel, OPEC (Organization of Petroleum Exporting Countries), the oil countries have taken over in varying degrees (from outright nationalization to majority share) the exploitation of oil and have increased substantially its price, apparently challenging the dominant position of the industrialized Western countries in the world economy.

Venezuela has been in the forefront of the action, promoting OPEC and strongly supporting a change in the terms of exchange between producer and consumer countries. Moreover, Venezuela has been a major supplier to the United States at a time when the United States is increasingly dependent on oil imports. In this global context, Venezuela's decision to nationalize U.S. and other foreign oil companies has a special significance—possibly dramatically altering global, hemispheric, and bilateral relations.

This study is divided into three parts. The first part analyzes the patterns of economic and social development within Venezuela leading up to the nationalization in order to cast some light on the nature of the social forces engaged in nationalization. By examining the socioeconomic forces that have emerged in the last decades, the scope and purposes of nationalization become clear. The rising and expanding indigenous capitalist class associated with an expanding and diversified U.S. industrial and service sector provides the necessary leadership, if not impetus, to nationalization. Under the influence of internal and external pressures, this burgeoisie has seized upon nationalization of oil to strengthen its own position in society, as well as to widen the opportunities for U.S. investment and trade while increasing regional opportunities for both.

The second section of the study is a detailed description and analysis of the plans and projects encompassed within the strategy of national capital expansion. It underlines the very narrow social stratum that will be the beneficiary of nationalization and posits the increasing integration of U.S. and Venezuelan industrial and banking capital, on the one side, and the increasing class polarization (between labor and capital), on the other. Thus, the growth of national ownership is seen to strengthen national capitalism and increasingly implicates it in a series of new relations of dependence with nonpetroleum foreign investors.

The third section examines the evolution of U.S. policy in relation to the structural changes taking place. The position adopted here is that while the aggregate of U.S. corporate interests provides the core of U.S. foreign policy, the state (through three of its key agencies) is the major instrumentality that interprets those interests and shapes policy. Through lengthy interviews and documentary analysis of the State, Defense, and Treasury departments, as well as interviews with corporate leaders, banking officials, and Congress, the main contours of U.S. policy are traced, identifying the high degree of consensus between corporate executives and the U.S. government's Executive branch on general policies and the sharp but limited areas of conflict over specific issues.

The study describes the Executive branch's early probing after a clear definition of Venezuelan elite attitudes toward such basic issues as private property, profits, foreign investment, trade, and the flow of capital. The crossing of the line between conflict and negotiation is defined early. The clear-cut and decisive commitment of Venezuela toward capitalist development in association with foreign capital paved the way toward a collaborative relationship between the United States and Venezuela. The study of the behavior of U.S. policy makers illustrates the fact that nationalization in the Third World is not perceived as such a revolutionary act as is sometimes believed. Once Washington recognized that nationalization in Venezuela did not affect the most dynamic sectors of U.S. capital, and, further, that it was a boon to U.S. exporters and investors and even to most of the oil companies that were expropriated, it was accepted.

While this account of Venezuela may shatter some of the myths about Third-World radicalism and challenges the rhetoric of a new world order, it also serves to point to a new set of alliances within and between countries: the salaried and wage workers in the advanced countries are paying the added costs while their counterparts in the Third World are receiving few of the benefits. It may be that some politicians and writers will describe the ties between the multinationals and the new rich industrial entrepreneurs (state and private) in the petroleum-producing countries as forming a "new order," but it is hardly a more equitable one and is not likely to be any more durable than previous hierarchies. In the postnationalization period, however, the struggles are less likely to be national and partial, but increasingly class-oriented and directed at the state, which now owns more but is just as exclusive as in the past in sharing the wealth.

This analysis is focused on the dynamic interplay of class forces within Venezuela and the reactions and policies emanating from the United States. Thus, the pitfalls of the so-called dependency school, which focuses exclusively on the impact of external factors and the constraints they impose on expansion, are hoped to be avoided. It was also sought to correct the narrow focus of the modernization theorists who treat countries and social forces in the periphery as autonomous actors, receiving or rejecting capital flows from inert external forces. The advantage of this approach is that it allows the interface of both internal and external political and social forces to be analyzed in an ongoing relationship.

One of the foremost concerns was to delve into the meaning of such historic acts as the nationalization of oil—what the policy implications are, both at home and abroad. By focusing the analysis on what governments do with their new revenues and earnings and how they affect citizens' lives, as well as corporate profits and international alignments, it is believed that one can determine some of the emerging tendencies that will influence the drift of history as well as the direction of U.S. policy. By identifying systemic imperatives, one hopes to clarify why governments, such as Allende's, which nationalized U.S. copper mines, are overthrown, while the United States accommodates and develops close ties with the Venezuelan regime under Pérez, which nationalized an apparently even more important resource—oil.

AN HISTORIC OVERVIEW OF THE VENEZUELAN SOCIAL FORMATION

To study any social formation is to study the way in which the essential classes that constitute it come into being. One of the crucial characteristics of the evolution of the Venezuelan social formation is the relative lateness of its crystallization—the long gestation period in the formulation of a bureaucratic-military apparatus compatible with the needs and desires of an emerging bourgeoisie. The history of Venezuela until the late 1950s was a period of personalistic military dictatorships, improvised economic activities, and a state apparatus that only partially articulated the essential needs of new productive forces. The political struggles of the second quarter of the twentieth century essentially were efforts by the newly emerging urban industrial forces to conquer and shape the state to their exclusive needs as a developing class. Having finally conquered the government in the late 1950s, the bourgeoisie set about to consolidate the state: to establish a firm grip on the military-coercive apparatus, to implant its ideology firmly throughout the society in order to foster the capitalist mode of production, and to establish routine channels for recruitment and disciplining of the labor force. The hope of the bourgeoisie was that once the capitalist state was consolidated it would preclude or severely limit the development of revolutionary politics.

The Venezuelan ruling class hoped to follow in the footsteps of its predecessors in Europe and North America. Yet the conditions in the latter areas were substantially different from those in Venezuela and the rest of the capitalist countries of the periphery. In the imperial West, consolidation of the state was achieved through the steady advance of the productive forces and the relative continuities in productive activity. In Venezuela and the rest of the periphery, consolidation has been infinitely more difficult. The implantation of the capitalist ideology is challenged from the very inception of the process of state building. Equally important, the fluctuations in the dependent economy, the abrupt shifts in external demand and prices, inhibit easy consolidation. At times lacking a secure and growing source of revenue to finance the establishment of stable and durable routines of civic society and at other times possessing funds but lacking social forces capable of channeling it into productive activity, efforts were largely concentrated in building up the repressive apparatus. The ups and downs of the economy and the shifts in productive activity made it difficult to discipline labor. Insecurity and a floating labor force generate a potentially explosive social movement, available to radical parties that promise full employment.

The transition from an unconsolidated to a consolidated capitalist state in Venezuela could be measured in two ways: the degree to which ruling-class dominance is exercised through ideological control (as opposed to physical force) and the extent to which the capitalist mode of production is implanted from the inside (rather than from the outside); only an internally anchored state is capable of institutionalizing the social relations of production necessary to reproduce a social formation that is both fully capitalist and national.

The nationalization of oil and iron was a major effort to consolidate the bourgeois state: to legitimize the bourgeoisie as an authentic national ruling class and to appropriate the funds to establish a diversified, internally controlled capitalist economy. The results of the nationalization to this point are inconclusive; indeed, the early returns are far from promising. The scope of external capital involvement continues to grow in precisely the areas of national diversification and this appears to be the intent of a bourgeoisie that has defaulted on its historic mission before it even got off the ground. The nationalization is in part, then, an effort at appropriation of the surplus with few historic tasks set and fewer resolved.

The failure of national capitalism to promote autonomous industrial development, even under optimal circumstances such as exist in Venezuela with its vast oil revenues, means that it is doomed throughout the Third World. The best that can be expected is an auxiliary role to foreign capital in a new mixed project in which, despite heavy inflows of revenues, external finance, technology, and management will play a dominant role. The part played by the national state is to collect funds, the national bourgeoisie channels the funds, and foreign capital directs and designs the major investment areas; together, they collect the profits. The long-term effects of this project are to unhinge the barely consolidated state, for it does not integrate labor in any positive position within the scheme of the new political economy. Most of labor is excluded by high capital intensity and the rest is exploited by the new concentrations of capital; under these circumstances the ideological levers have little effectiveness in generating support, leaving only the use of force as the only durable weapon for sustaining the regime. Hence, while the oil nationalization theoretically could have served as the instrument to consolidate the capitalist state, it has in fact set in motion a set of contradictions that can, in time, unravel the entire social and political order.

THE IMPERIAL STATE AND FOREIGN POLICY

The imperial state is a direct outgrowth of the competitive period of capitalism and coincides with the growth of monopoly capitalism. The imperial state emerged as a necessary accompaniment of capitalist expansion on a world scale. V. I. Lenin's writing on imperialism, necessarily confined (by the czarist censor) to the economic dimensions of the phenomenon, avoided any substantive discussion of the role of the imperial state. Nicolai Bukharin, in his work on imperialism, sketches out some of the activities of the imperial state:

> The state apparatus has always served as a tool in the hands of the ruling classes of its country, and it has always acted as their "defender and protector" in the world market; at no time, however, did it have the colossal importance that it has in the epoch of finance capital and imperialist politics. With the formation of state capitalist trusts, competition is being almost entirely shifted to foreign countries; obviously, the organs of the struggle that is to be waged abroad, primarily state power, must therefore grow tremendously.[1]

He then proceeds to enumerate imperial state activities, including tariffs, protection, subsidies, commercial treaties, loan guarantees, and military intervention. Bukharin's perception of the growing importance of the imperial state was correct, yet he did not fully elaborate on the far-reaching efforts and regular functions that which state would perform with the passing of time. Moreover, to no small degree, Bukharin exaggerated, on the basis of the particular experiences of World War I, the degree of fusion between the state and capitalist enterprise and hence drew his correct conclusions about the growth of state involvement on the basis of a wrongheaded diagnosis of the relationship between state and capital:

> ... the future belongs to economic forms that are close to state capitalism. This further evolution of the state capitalist trusts, highly accelerated by the war, is reflected, in its turn, in the worldwide struggle among state capitalist trusts.[2]

The imperial state emerges from Bukharin's analysis of the fusion of the state and capitalism during World War I and is projected forward in time and extended throughout the world—a path-breaking effort, but one that fails to elaborate clearly the specific roles of the state in the different phases of capital accumulation and that does not clearly delimit the state from capital and the specific role one plays in relation to the other.

The cyclical nature of capitalist expansion and the reverses suffered through political and military struggles and economic rivalries temporarily diminished the relative importance and involvement of the imperial states during the interwar period, preparing the way for the massive confrontations that were unleashed during the European and Asian bloodbaths of the 1940s. Unfortunately, both the ups and downs of imperial state activity failed to generate a significant body of literature on the imperial state. In the United States, the crude approximations of a discussion of the imperial state were found in the writings on dollar diplomacy, the big stick, and gunboat diplomacy, which touched upon epiphenomenal occurrences, actions that sustained particular policies during conjunctural moments. The marines were still conceived of as an isolated force safeguarding particular interests rather than as the institutional foundation of a new stage in capitalist development.

The revival of capitalism after the 1930s and the subsequent wars of conquest, colonial and neocolonial wars in Europe and the Third World, provided a substantive basis for a reconsideration of the role of the capitalist state as part of a global system of exploitation. This task, nevertheless, has not been confronted. At best, theoretical and some empirical research have started to examine the capitalist state within the boundaries of the national state—unfortunately at a time when the prime basis and focus of the state is no longer so confined. The weakness of this new research on the capitalist state is evident when one attempts to project outward the inner activities of the capitalist state. The imperial state is not the capitalist state writ large. While there appears to be some repetition of function at the international level, the scope and interests involved do not allow for analogies, which are at a loss to explain the substantial disparities that are accumulating. The differences between the capitalist and imperial states cannot simply be reduced to a contrast between, as Franz Fanon would have it, the democracy of the metropolis and the colonial state in the periphery, between ideological and violent forms of domination. The differences in form of rule are submerged within the same state, exercising its role as the central instrumentality of exploitation, concentrating its force and ideological weapons on a single unified project: the accumulation of capital on a world scale.

Worldwide expansion of capitalism concentrated in its great national centers has made possible and, in turn, necessitates the growth of a state that formulates its strategy and policy on a world scale. The foundations of the imperial state are grounded in the world division of labor. The appropriation of the surplus on a world scale is the central task of the imperial state and this requires the elaboration of a new set of machinery of administration, as well as a new division of powers and delegation of responsibility based on global calculations. The imperial state, as the U.S. "external budget" indicates (through military spending, loans, credits, agency allocations, CIA funding, and so on), has grown far beyond the needs of an internally directed nationalist capitalist political economy.

The historic progression of capital on a world scale has been made possible only by commensurate activity by the imperial state. Accumulated capital does not find its own direction or outlet, but can solve the realization problem on a global basis only by the direct intervention of the imperial state. The creation of the imperial state is not a product of an act harnessing romantic dreams of imperial grandeur to the prosaic economic needs of particular entrepreneurs. Nor is it the product of bureaucratic imperatives or psychological drives. Rather, the imperial state reflects the cumulative extension of bureaucratic agencies around the imperatives induced by the dynamic of capitalist expansion. After the initial military forays, trading companies, and state loans come the formal institutional networks of banks and monetary systems. The bureaucratic increments in overseas activities are, at a certain point, qualitatively transformed into large-scale, full-time institutional configurations whose capacity to sustain activity is no longer subject to the calculus of individual firms and their internal needs. Today, the bulk of the major corporate and banking units depends on external exploitation to grow and survive.

As the axis of the economic system turns essentially on matters of global import, the functioning and structure of the state turn on its capacity to create parallel institutions that can sustain the new economic foundations. The close ties between imperial state and economy vitiate any attempt to seek initial causes or primary determinants. It is impossible to conceive of an imperial economy without an imperial state. The reciprocal nature of the relationship exists from the point of entry into the world market through the consolidation and continued reproduction of imperial social relations to the struggle for survival that constantly emerges as a consequence of the implantation of exploitative relationships.

In the abstract, what informs the imperial state are the central conditions of capital accumulation, the organization of labor, and the availability of resources for exploitation within a social order capable of sustaining itself over time. Domination is buttressed by the use of violence in all of its forms, ranging from direct and massive assaults and selective terror to ideological mastery. The imperialist state has generalized what Nicos Poulantzas describes as the "exceptional" state (fascism, Bonapartism, military dictatorship) as the "normal' state. The economic project of the imperial state tends to eliminate socially significant strata which "mediate" the class struggle, leaving mainly the repressive apparatus as the major source of sustenance. However, the political function (the fortification of the repressive organs in the periphery) of the imperial state is merely a moment in the process of imperial expansion; during the process of creating the foundations of the new world division of labor, the economic function becomes central.

Concretely, the imperial state drives economic expansion forward but does so only at the behest of specific factions of the imperial bourgeoisie. The particular ties between imperial capital and the peripheral economy and society shape and direct the particular policies and strategies that the imperial state adopts in the conjuncture. While the imperial state moves according to the imperatives of the imperial system, it does so only within the parameters set by the dominant capitalist forces operating within the arena. Within the imperial economy and state are the hegemonic capitalist forces laying claims to resources and commanding influence. It is within these two areas—systemic imperatives and social forces—that the specific policies of the imperial state can be analyzed.

The imperial state involves the whole ensemble of institutional forces that operate throughout the world and whose primary concern is shaping the conditions for accumulation on a world scale. From the Treasury and State departments to the Pentagon and the World Bank, the imperial state embodies the basic notion that the primacy of U.S. capital is the prior condition for the growth of the world economy. A loss of power, as occurs when a revolution turns a region out of the orbit of U.S. domination (as in Vietnam), is, according to this standard, a measure of the failure of the world political order. The hegemonic position of the United States within the world order is the sine qua non for the imperial state. In the world of nation-states with competing modes of production and economic rivalries, the imperatives of preeminence are adapted to a multiplicity of circumstances, which are not always within the control of the imperial state. The process of devising a policy in a given area revolves around the strategic devises of specific agencies and decision makers with their own institutional interests and idiosyncratic dispositions.

The imperial state's functions and structures cannot be deduced by examining intra-agency disputes and conflicts, as is proposed by the theorists of bureaucratic politics. However, a situationally rooted study allows one to formulate a partial view of the manner in which the decision makers within the agencies of the imperial state formulate policy. The determination of criteria that inform action—the position of the imperial state in a particular conflict—provides close-up knowledge of the operational meanings of imperialism in moments of crises. On the other hand, abstract Marxists, whose knowledge of imperialism as a social system is based on the exegesis of functions and instrumentalities, fail to identify the concrete operative principles that inform imperial action. Abstract Marxism, devoid of operational meaning, leads to a multiplication of taxonomic categories and propositional statements that then are illustrated by an arbitrary selection of historic facts. The lack of scientific rigor in this rather mechanical exercise leads to endless ideological confrontations that, on their own terms, cannot be resolved: it remains a confrontation of positions with the appropriate illustrations. Thus, it falls on empirical analysis to bring out the specific features that identify the imperial state and give meaning to its behavior. Otherwise reality is stood on its head and abstract theory is considered a way to resolve the imperial-induced crises of our time.

NOTES

1. Nicolai Bukharin, *Imperialism and World Economy* (New York: Howard Fertig, 1966), p. 124.
2. Ibid., p. 158.

The Nationalization of
Venezuelan Oil

CHAPTER 1

CAPITALIST DEVELOPMENT IN HISTORIC PERSPECTIVE

INTRODUCTION

In order to understand the meaning of the 1976 nationalization of petroleum by the government of President Carlos Andrés Pérez, it is useful to try to put the event in its historicl context. As a general format for doing so, three areas have been abstracted from Venezuela's past: the development of a capitalist mode of production, the formation of that mode's social classes, and the evolution of a capitalist state. Within these three broad areas of concern, attention is focused on the growth of a home market, the formation of the Venezuelan financial oligarchy, the development of a capitalist state structure with agencies and procedures suitable to the interests of a national bourgeoisie, and the particular policies of succeeding governments that have furthered the formation of that class.

Within each of the periods studied, it is shown how, as commodity exchange and capitalist relations of production develop, they give rise to specific social classes. These classes provide both a constellation of material interests and the collective and individual class agents (social forces) to further those interests. The Venezuelan ruling class, due to its dependent association with foreign capital, has historically been more reliant on the state to advance its interests than have ruling classes in other countries. Because Venezuelan governments also have devised particular policies to respond to the perceived objective needs of the capital accumulation process at given moments, social forces from other classes have often acted through the state on behalf of the ruling class, members of which have occasionally balked at the very measures whose long-term effects have benefited them.

This account illustrates the process of capitalist transformation, which, in the Venezuelan context, has occurred without the benefit of a bourgeois-democratic revolution, without a violent attack on landlords or imperialism. The contrary appears to be the case: a closer association among the industrial, commercial, and agricultural capitalists and between the North American and Venezuelan bourgeoisies as a whole. The political regimes that incorporated the capitalist mode of production have included military dictatorships and parliamentary governments. Indeed, as this historic survey illustrates, there was substantial continuity between the Pérez Jiménez regime, on the one hand, and the AD (Democratic Action Party) and COPEI (Christian Democratic Party) governments, on the other. The transformation of Venezuela from a precapitalist to a capitalist social formation having initially an agrocommerical and later a mineral-manufacturing and service economy has not been accompanied by egalitarian socioeconomic changes. The major result has been the emergence of an industrial bourgeoisie tied to state and foreign capital, its hegemony based on lateral alliances rather than on mobilization from below. The growth of industrial capitalism has been, and continues to be, an elitist and exploitative phenomenon that contains few, if any, of the progressive attributes often associated with the rise to power of the bourgeoisie.

LEGACY OF THE IMPERIALISM OF FREE TRADE

By the time Royal Dutch Shell sank the first oil well in Venezuela in 1914, agricultural exports, the previous leading sector of the economy, were already moribund (although this was obscured for a time by the Brazilian governments' stockpiling of Brazilian-grown coffee). However, the class structure and political formations of the agricultural period would condition the policies of subsequent Venezuelan governments as much as would the evolving location of the nation within the world capitalist system.

During the last third of the nineteenth century, when 80 percent of Venezuelan exports were in coffee and cacao, Europe was already at the end of its period of original accumulation and was producing and exporting both real and money capital. Europe's relatively well-developed commodity exchange and capitalist production relations were leading to the appearance of accentuated business cycles, which greatly affected its demand for those consumer goods imports that happened to be easily substitutable, such as coffee and cacao. For Venezuelan export-related productive activity, this was experienced as wild fluctuations in revenue.

Because the Venezuelan productive forces were as yet too undeveloped to have allowed many large accumulations of wealth, and because it was essential for primary product exporters to have access to the European market outlets, the major import export merchants of Maracaibo and Caracas were largely foreign.[1] In effect, these foreign owned mercantile houses came to function as the nation's central bank and treasury. Venezuela's money supply was determined by how much merchandise the German, French, English, and U.S. (beginning in World War I) metropolitan headquarters would advance their affiliates in Venezuela. Such merchants, who made most of their profits from the sale of imported manufactures rather than from agricultural exports, determined the state of commercial agriculture by controlling the interest rates and the acceptable levels of accounts payable of their retail outlets in the interior. Also the relative prices that they set between consumer and capital goods imports greatly influenced the structure and rates of productive investment. The greatest share of the economic surplus produced by Venezuelan labor was not available for national reinvestment as it was appropriated by the foreign mercantile houses either via the terms of foreign trade or profit remittance abroad.[2]

During the first decade of the twentieth century the profitability of agricultural exports declined. However, the governments of first Cipriano Castro and then Juan Vicente Gómez were severely constrained in any efforts either to change the nation's status as a primary product exporter or to develop a home market through implementing import-substitution measures. The extremely heavy foreign debt burden incurred by the Castro government* combined with the continuing stagnation of coffee exports custom duties caused a series of defaults on the loans. The nation effectively went into financial receivership following an invasion by the armed forces of the creditor nations. The outstanding loans, being payable only in the same foreign currencies, required the continued exportation of those primary products demanded in Europe. World War I's disruption of traditional European-Venezuelan trade and credit ties provided some opportunities for spontaneous import substitution in such areas as textiles, beer, and rum, construction materials, and electricity generation. Two of the nation's current ruling families, the Vollmers and the Zuloagas, combined in 1910 to form the Caracas Electricity Company, among other ventures, and were able to prosper during the period of the interimperialist war. Still, the possibilities for a domestic market were rather narrowly circumscribed by the limited size of both the urban and rural proletariat and by the total lack of transportation between the export enclave regions; the cattle plains, which were outside the market altogether, continued to retain over one third of the nation's population in 1920.[3]

As the Venezuelan landlords began to seek alternative areas in which to invest their wealth,[4] they found that many of the scarce opportunities had already been preempted. The private credit market and foreign trade were largely in the hands of European merchants, and mining, the single most

---

*Total public debt, most of which was held by foreign capital, was equal to 40 percent of the gross national product in 1900.

profitable activity in the economy, was monoplized by one company, El Callao. Many of the national landed and commercial gentry—typified by the Boultons, an old, English coffee-exporting family of Caracas who had reached its apogee of class power by financing the Guzmán Blanco government's suppression of the civil wars of the 1890s—turned to the provision of public credit. In 1890, there were three major Venezuelan banks: the Bank of Venezuela and the Bank of Caracas, both subscribed by the cacao and merchant bourgeoisie of Caracas, and the Bank of Maracaibo, which was underwritten by the coffee interest of the Andean-Maracaibo region. As the following table shows,[5] by 1920, the Bank of Venezuela had become the preferred outlet for the wealth of the national bourgeoisie and planter classes due to its favored position vis-a-vis the Venezuelan state. [6]

Family	Percent
Boulton	30
Eraso	30
Rohl	20
Santana	10
Calixto Leon	10

However, even more than public debt it proved to be urban construction, which offered high profits and required little entrepreneurial ability, that was clearly the beneficiary of the decapitalization of agriculture prior to the oil boom.[7]

Several features of the national social formation that were the legacy of the periods of colonialism and free trade imperialism would have important effects on the subsequent development of Venezuelan capitalism. While the hegemony of foreign capital over the nation was not always experienced directly by the Venezuelan masses, the reality of foreign domination provided the basis for a nationalist ideology that would later be exploited by all the major parties in their electoral campaigns. The national bourgeoisie was never progressive. It originated from the convergence of planter and merchant wealth in the banks created at the turn of the century, which would provide the cornerstone for the formation of the Venezuelan finance capital groups that emerged after World War II. These banks, which cornered the market in urban real estate prior to the reign of oil, would later be in a position to establish a virtual monopoly over industrial credit. Because the national bourgeoisie was a monoply bourgeoisie from the outset, its primary interest was always in valorizing existing capital, rather than expanding production or reducing its cost (to the extent it invested in productive capital at all).

Also due to the hegemony of foreign capital in the agricultural period, the national bourgeoisie would remain a dependent class in later years, compelled either to associate with foreign capital or cultivate ties to the Venezuelan state.

The combination of an undeveloped home market incapable of integrating the social formation, the recurrent instability of the export market, and a regionally divided ruling class led to considerable political instability. In attempts to solve the political problem there developed a tradition of reliance on a relatively autonomous militarist state under the leadership of a succession of regional caudillos (military dictators). This institution would last well into the period of direct foreign investment in petroleum, with the national bourgeoisie only really coming to dominate the state in the 1960s.

MILITARY MERCANTILISM

By the time Presidnet Vicente Gómez died in office in 1935, petroleum had become the pivotal sector of the economy. Between 1925 and 1936, the value of oil production increased 835 percent (having surpassed agriculture as the largest sectoral contribution to gross national product in 1927[8]), while the nonoil sector grew 156 percent. Within the latter, artisan production and industry both expanded 124 percent (largely in those traditional mass consumption goods that did not compete with foreign imports, for example, rough textiles and sandals), commerce and services increased 109 percent, and commercial agriculture trailed at 39 percent.[9]

Despite the increase in government revenue and national income resulting from British and North American investment in Petroleum,[10] this growth had a negative impact on the productive structure of the Venezuelan social formation. In general, the foreign-owned oil export sector rapidly became an enclave, much more disarticulated from the rest of the economy than commercial agriculture had ever been. This became readily apparent with the U.S. stock market crisis of 1929. As a result of the crash, the American petroleum corporations reduced employment of Venezuelan labor by one third—an indication that levels of production within the oil enclave were determined independently of any decisions made within the Venezuelan economy.

Besides the sectoral disarticulation there developed by 1935 a disjuncture between national production and investment. Because of the sudden rise in national income, coupled with the preceding decay of export agriculture, a massive transfer of capital and considerable labor migration occurred toward the urban centers from the six states (Táchira, Trujillo, Lara, Sucre, Miranda, and Mérida) that had previously produced the largest share of the nation's coffee exports. The rural oligarchy behaved exactly as any modern capitalist entrepreneur. That is, within the constraints set by the existing configuration of available investment opportunities, it sought the projects in which the anticipated profits were greatest relative to the risks assumed. In the context of pre-World War II Venezuela—fragmented markets, undeveloped productive forces in construction materials and electricity, a small and unskilled urban proletariat, and a highly skewed national income distribution—the most attractive investments continued to be in real estate speculation, construction, and commerce.[11] That expanded output in artisan and industrial production that did occur during the 1920s and 1930s, rather than being a response to new investment stimulated by the growing export of oil, was the result of the full utilization of the earlier investments that had been brought on by World War I.

Another disjuncture created by the foreign investment in petroleum occurred between national production and employment. The initial construction of the infrastructure required for the extraction of crude oil provided a source of employment for the displaced rural coffee workers. However, the actual production process utilized an import capital-intensive technology. Between 1920 and 1940, the value of imported capital goods for the oil sector was more than twice the amount spent on labor.[12] As a result, the overall employment level of 1929 was not reached again until 1946, despite the fact that oil-related national income increased 35 percent from 1929 to 1940.[13] On the other hand, even though commercial agricultural production had dropped to 18 percent of gross national product by the time Gómez died in office, the nation's population was still more than two thirds rural and 58.2 percent of the work force was marginally engaged in agriculture.[14]

The exploitative nature of the foreign oil investments manifested itself after 1927. Coincident with the establishment of the industry's fixed capital assets, profits plus depreciation exceeded new foreign investment as retained earnings allowed the corporations to become self-financing. In addition to the foreign consumption of Venezuela's most valuable primary resource, the nation became from then on a net exporter of capital.

In spite of the uneven development of the coastal enclave and the rural sector, it would be wrong to conceptualize the two sectors as constituting a dual society. Both continued to be two parts of a single system of production oriented toward the appropriation of the nation's economic surplus by foreign investors accumulating capital on a world scale. Agriculture had generated the initial foreign exchange and personal wealth upon which the coastal towns had been built. Although in this period increasing shares of the nation's food supply and industrial raw materials were imported, the internal terms of trade between food and manufactures supplied another lever of original accumulation for the urban merchants. Finally, the rural sector provided a latent reserve army of labor for both the construction of governments' public works and the oil corporations' infrastructure.

At the level of the formation of social classes, one of the most striking aspects of the development of Venezuelan capitalism has been the telescoping of what were, in the experience of the advanced capitalist nations, century-long processes, and the inversion of the normal sequence of class emergence. The period between 1920 and 1935 witnessed the rapid development of both propertied and salaried urban petty bourgeoisies. The former consisted of self-employed artisans who migrated from the declining rural areas, attracted by the possibilities of employment arising from the accumulation of oil-related wealth in the cities. Despite a lack of capital, this class contributed substantially for a time to national production. Because of the structural deficiencies of capitalist production in the nonoil sector, their workshops served as a stopgap prior to the establishment of factory production. Numerically, these small manufacturers reached their zenith in 1936, growing from 43,000 in 1920 to 126,000 (27.3 percent of the nonagricultural work force), a level at which they would remain until 1950.[15] After the late 1930s, this class's social weight would be considerably reduced, as the combination of foreign import competition and the changing of consumer tastes away from traditional artisan products (resulting from the continued urban migration) would erode its ability to affect production.

Venezuela's history of foreign domination imposed upon it a particular location within the world capitalist system. The result for the nation was the creation of an economy with three major abiding characterisitics: the concentration of nationally owned wealth into largely unproductive activities, the predominance of foreign-owned capital within an industry using a labor-saving technology incapable of absorbing the flow of rural migrants, and the existence of a state enjoying a claim to a large portion of the national income due to its ability to tax the oil corporations.

The result for the national class structure was the formation of a large salaried petty bourgeoisie that developed prior to a national industrial proletariat. One stratum of this class worked in the private sector (commerce, banking, and oil-related services); a second was made up of the mushrooming state bureaucracy, which expanded from 13,500 to 56,100 between 1920 and 1936.[16] Overall, this class experienced continued growth after the passing of the Gómez regime, increasing from 37.9 to 54 percent of the nonagricultural work force by 1950[17]—a fact that would have an important impact on subsequent political behavior and institutions of the country.

By 1930, the transition from agricultural exports to petroleum as the leading economic sector had an impact on political arrangements. The president's cabinet was restructured to better accommodate itself to the social formation's newly arising needs. The Department of Labor and Communications was established in 1937 and a separate Department of Labor was created in 1945. By 1950, a new Ministry of Mines and Hydrocarbons was formed. Within the cabinet during the Gomecista years, there were two groups represented: fellow military men from Táchira and the Caracas merchants, the latter increasing in predominance by the mid-1930s. In general, the state was highly centralized and strongly militarist; even the state governors were military appointees. Only the propertied classes were allowed to vote, and, in 1928, elements from the growing number of salaried professionals and university students began agitating for broadening the franchise.

In the fashion typical of neocolonial military dictatorships, Gómez' general policy orientation was one of laissez-faire economics and political authoritarianism. Ruling in the interest of domestic merchant and bank capital, the foreign oil companies, and military, the regime created deflationary budget surpluses, decreed regressive taxation policies, encouraged free importation, and maintained a stable currency with free convertibility to foreign exchange (to allow easy repatriation of profits). After 1920, the state replaced foreign capital in assuming the cost in building an economic infrastructure: first in roads, ports, and buildings and after 1938, in the construction of public housing and the provision of investment credits.

During the late 1920s and early 1930s, the Gómez government in effect financed the transition of the rural oligarchy from agricultural export activity to speculation in urban real estate. The planters, faced with declining opportunities for profit taking in commercial farming, took out government mortgages on their rural properties from the Agricultural and Livestock Bank, applied the principal to their urban activities, and then defaulted on the mortgages—allowing the state bank to foreclose and assume the titles to their increasingly worthless lands, for which there would have been no buyers.

The influence of the urban commercial bourgeoisie and the *latifundista* on government policy was evident in the Tinoco Convention (1934). The measure established government control of exchange rates and the money and credit supply via the Agricultural and Livestock Bank and the Centralized Office of Exchange (the precursor of the Central Bank of Venezuela). The main effect of the policy was to break up the monopoly of the credit supply that the German import-export houses still maintained and to open up financial channels for the emerging local bourgeoisie.

Regarding the government's policy toward the foreign oil interests, Gómez was content to provide the public treasury with a steady supply of revenue while accumulating a huge family fortune in the process and doling out land to his political loyalists. From 1919 to 1935, the estimated earnings of the petroleum corporations were taxed at a modest 7 percent, while government receipts were augmented by the sale of additional offshore concessions.

THE FAILURE OF NATIONAL POPULISM

Capitalist Transformation and the Ascendency of the Petty Bourgoisie

Between the end of the Gómez period and the overthrow of the *trienio* regime of Acción Democrática in 1948, the process of original accumulation gradually became one of capital accumulation proper, as the concentrations of merchant and bank wealth were transformed into industrial assets based on capitalist relations of production. The world capitalist crisis, followed by World War II, gave the Venezuelan bourgeoisie another opportunity for spontaneous import substitution. In addition, the U.S. and British governments, because of their acute need for increased amounts of crude oil to fuel the Allies' war machine, did not try to prevent the Venezuelan governments of the period from significantly increasing their share of the oil revenue.[18] Added revenues allowed the state to increase its subsidies to national capital and to continue in its role as a major employer, both of which contributed to the expansion of the home market.[19]

Manufacturing began to expand rapidly but selectively after 1945. Between 1945 and 1950, industrial growth occurred only in those few areas to which the state gave tariff protection: food processing, beer, artificial silk, and some construction materials.[20] New construction techniques and the importation of electric generators after the war provided the physical capacity and power source for factory production. The increasing commercialization of agriculture resulted from the increasing demand from the accelerating urbanization of the 1940s (see the following table[21]), the increasing income of the oil workers and salaried petty bourgeoisie, and the influx of state aid to rural areas and investors during the trienio period (1945 to 1948):

Year	Annual Rate (Percent)	Urban Population (Percent)
1920	—	26.1
1936-41	0.5	31.3
1941-50	1.8	47.9
1950-61	1.3	62.5
1961-70	0.7	70.0
1970-75	2.0	80.0

All stimulated capitalist production in such foods as milk, meat, eggs, and vegetables. After 1950, as industrial monopoly capital began to engage in vertical integration, sectors of agriculture devoted to industrial raw materials would surpass food crops in economic importance.

In spite of the more nationalist trienio oil policy and its more effective promotion of national industry, foreign (primarily U.S.) direct investment in Venezuelan industry increased substantially after the war.[22] U.S. capital, for example, moved into Venezuelan electric utilities, which doubled in number within those three years. Even in cases where controlling interest was held by Venezuelan banks, the banks themselves were often owned predominantly by foreign capital. The same was also true for the Agricultural and Livestock Development Bank, which provided much of the commerical credit to tobacco, cotton, and milk farmers in the 1940s and 1950s; the Cavendes Bank, which underwrote the sale of stock in the former, was controlled by large domestic and foreign capital.

In the 1935–1948 period, the major activities for capital accumulation were the middle level service industry, light manufacturing, and real estate and commerce. The original ruling merchants and bankers had begun to organize themselves into finance capital* groups by the end of the war in order to pool the capital, technology, and political influence of others. The nature of this ruling financial oligarchy began at this time to emerge in three characteristic ways. First, it was much less directly integrated with the oil export enclave than had been the case with the earlier agricultural exports sector. Second, it came to be increasingly dependent on various types of state subsidies based on the channeling of public revenue derived from the foreign-owned petroleum sector. And third, it became ever more closely tied to foreign capital outside the oil enclave. These features were particularly evident in the case of the three kinship groups that dominated this period of transition from mercantile to industrial capitalism.

The Vollmer family, as the most important shareholder in the Commercial and Agricultural Bank and the nation's biggest real estate speculator, was the chief beneficiary of the governments' policy of keeping public revenue on deposit in the private banks. One of the first industrializing groups, it established itself as a vertically integrated monopoly by investing in sugar plantations in order to supply its rum distilleries.

The Boulton family empire had suffered in the 1920s from the declining coffee prices and from the loss of markets, as the United States replaced England as Venezuela's principal trading partner. In 1929, the family turned to the Venezuelan state for rescue from financial disaster. The Gómez government bought several of its urban properties in order to prevent a national chain reaction of bankruptcies. Since the 1930s, the family has concentrated its investments in the importation of luxury consumer goods, oil-industry-related services, and manufacturing. The Boulton group benefited from its direct ties with the Medina Angarita regime via the president, who enabled it to secure import licenses. The Boultons have been closely tied to foreign capital as well, since the family became the shipping agent for both Grace Lines and Pan American Airlines.

Eugenio Mendoza, today among the most powerful individuals in Venezuela, began from relatively humble origins in the 1930s as an importer of artisans' tools and hardware who also speculated in urban property. After Gómez' death he was favored in government subsidies, monetary policies, and in the award of government contracts for supplying public works projects with

---

*Throughout this work, the term "bank capital" will refer to owners-units of all types of fictitious (money) capital, not just commercial banks; the concept "finance capital" will be employed in the Leninist sense, that is, owners-units of combined bank and industrial (real) capital.[23]

construction materials. Mendoza also came to be very closely associated with foreign capital. His cement and construction materials firm, Vencemos, enhanced its monopoly position considerably in 1948 when Mendoza got the patent rights from the North American Sherwin-Williams Company, a paint manufacturer, and became that company's sole Venezuelan representative. In 1944, Mendoza formed the first Venezuelan agribusiness corporation, *Protinal,* which invested in branches of the intensive food industry, such as eggs, milk, and chicken—displacing products formerly imported from the United States —by affiliating with English and Japanese capital.

The political cohesiveness of the bourgeoisie was increased in 1944 with the establishment by the chambers of commerce of Caracas and La Guaira of the Venezuelan Federation of Chambers and Associations of Commerce and Production (Fedecámaras), a peak organization of business interest groups composed of small and medium-sized industrialists and merchants. Its initial purpose was to unify Venezuelan businessmen as a defense against the labor union confederation formed under the leadership of AD. It served the function of rationalizing and managing rivalries among the competing units of capital while offering an alternative vehicle for access to the state for those who lacked personal ties with members of the particular governments.

Both the social weight and the organizational coherence of the industrial proletariat increased between 1935 and 1948. The latter was chiefly the result of the growth of the labor union movement (particularly effective in the oil sector), whose capacity for mobilizing the working class politically was enhanced by the aggregation of larger numbers of laborers into fewer units of production. The working class's social weight increased apace with the improvements in industrial productivity. In this respect, however, increases in the amount of fixed capital per firm mattered more than the number of employees per firm, as the newer production technology was generally capital intensive. From 1941 to 1953, the number of industrial workers actually declined from 173,046 to 167,726.[24] This absolute net loss of industrial employment was the result of decreases in the number of artisan workers that more than offset the increases in factory employment. This tendency in the changing social relations of production, combined with extremely rapid urban migration, together were responsible for the swelling of the ranks of the nation's industrial reserve army.

The salaried petty bourgeoisie, supported by the fast-growing oil revenue, became even more important than before due to a wave of postwar foreign immigration. The economic significance of this class was clearly expressed by the meteoric rise of the Union Bank. Today the second largest bank in Venezuela, it was founded at the end of World War II by a group of small immigrant merchants along with a Venezuelan family whose wealth had been first accumulated in cacao and bananas and then cattle and rail transport. At the time it was incorporated, all the other banks were tied to the government or relied on personal ties to monopoly capitalist subscribers. The Union Bank sought, and got, the small savings and checking deposits of the numerous salaried petty bourgeoisie and provided them with access to credit. It specialized in consumer financing of those imported durable goods demanded after the war and, by 1945, had already become the fastest growing enterprise in Venezuela. Because this class was able to provide political leadership in the AD party capable of mobilizing large numbers of peasants and workers over bourgeois democratic issues, such as direct elections, universal suffrage, civilian representative government, agrarian land reform, and so on, it also became a political force with which to be reckoned.

The major contradiction within the Venezuelan state in this period was located in the military. The contending forces were the old guard Gomecistas versus the new technicians of the officers corps, and the major issues concerned inequities in levels of pay and rank advancement. In 1945, the dissatisfied junior officers organized the Patriotic Military Union, overthrew President Medina Angarita, and then invited the AD to organize a civilian government.

The three-year trienio government that followed was essentially nationalist-populist. The party leadership was solidly from the rising salaried petty bourgeoisie, its foreign policy was aggressively nationalist (within the limits set by the extent of the nation's capitalist development), its electoral base was polyclass, and its economic policy was designed to provide social welfare while stimulating capitalist growth.

The demands of the military, which had precipitated the coup, were dealt with in 1945 during the interim government of the junta. Decrees were issued that instituted professional officers training programs, established the appointment of civilian state governors, purged the army of the leadership of the Andean regional clique-personality cult, and in general demilitarized the state apparatus.[25]

One of the AD's most popular party platforms was implemented within a year of the coup: the abrogation of the 1936 constitution, which had effectively excluded the vast majority of the population from participation in the parliamentary elections. The government advanced the interest of the university-trained professionals by greatly increasing the size of the state bureaucracy and by maintaining the privileged legal status of "nonmanual" employees. Labor unions, particularly in the oil sector, were allowed to press employers for wage increases, but under the auspices of the Confederation of Venezuelan Workers (CTV), which had been established by the Medina Angarita regime in 1944 as a way of purging the dominant Communist leadership in the trade union movement.

The *adeco* government created a similarly clientelist and economistic peasant league, the Federation of Venezuelan Peasants (FCV), dominated by AD leadership in 1946, and began to implement an agrarian reform program based on the modernization of production and the colonization of government land. Education was taken out of the hands of the Catholic Church and then extended and modernized. Government oil revenues were greatly increased and made increasingly available to the bourgeoisie, and the economic infrastructure, especially agricultural irrigation and electric power generation, was expanded. The social and electoral reforms of the provisional government made AD immensely popular, as was evident in 1948 when the party's presidential candidate, Rómulo Gallegos, received 74 percent of the popular vote.

Oil Policy and Venezuelan Development

The primary and abiding concern of Venezuelan governments since the presidency of Gómez' successor, López Contreras, had been how to guarantee the flow of state oil revenue to state enterprise and private national capital in order to underwrite the industrialization process, while minimizing the problems associated with remaining an export enclave dependent on foreign capital.

The doctrinal basis of Venezuela's oil policy since 1937 was expressed by López Contreras, who described the use of public oil money to accelerate capitalist development within Venezuela as "sowing the oil." López, whose hydro-carbons law established the first unified government policy vis-a-vis the foreign oil corporations as a whole (in place of separately negotiated concessions), was able to exploit the vulnerability of the U.S. government after 1938 to increase the tax on exported crude oil from 7 to 20 percent.

It would be difficult to mistake either López Contreras or Medina Angarita for ardent nationalists, however. During their combined tenures in office, the oil companies were not prosecuted for tax evasion,[26] previously held oil concessions were extended 40 years, and huge new concessions were sold —equal to more than all the earlier ones combined. Several of Medina Angarita's ministers had direct ties with oil corporations,[27] and the 1943 law establishing that regime's oil policy, although it was instigated by an AD congressional deputy, Pérez Alfonzo, was actually written by the U. S. oil industry experts working with the government's attorney general.[28]

After the 1945 coup, Pérez Alfonzo became minister of the newly formed Venezuelan Development Corporation (CVF), which was responsible for oil policy until 1950. Under his administration the tax rate per barrel was raised to 50 percent and limitations were placed on the further sale of concessions. A government oil corporation, the Venezuelan Petroleum Corporation (CVP), was created that established contractual relations regarding the oil companies' use of Venezuelan subsoil and that served as the embryo for a national mixed company for the refining of oil. The government began receiving the royalty payments of the foreign oil firms in crude oil, which the public corporation could then itself market abroad. To reduce the state's yearly budgetary dependence on oil and to counter the foreign capitalists' coercive manipulations of the price of oil and the levels of production and investment, an "antibusiness cycle fund" of reserve oil revenues was established.

Given the availability of the economic surplus in the hands of the state and a large bureaucracy capable of administering its disposition, there were alternative ways the post-Gómez governments could have stimulated national capitalist development—for instance, by encouraging several weak firms to merge or by relying on state enterprise. State subsidies and credits to capitalists were selected as the preferred methods because of the preexistence of bank, industrial, and commercial monopolies and the absence of a large, militant, and organized working class enjoying independent leadership. Due to the large increase in oil revenues during the late 1930s and the 1940s, the state came to be more important in the financing of Venezuelan industrialization than the other previous sources, that is, commercial and urban land speculation.[29] After 1936, the state devoted more money (both in absolute terms and as a share of the federal budget) to the development of the economic infrastructure. In addition to the continuing construction of roads and government storage facilities, irrigation and improved public sanitation were also emphasized.

Even more striking, particularly during the brief AD regime, was the direct financial aid given to capitalist agricultural and industrial investors. In addition to the establishment of the CVF in 1946, there were its predecessors, the Venezuelan Industrial Bank (BIV), founded in 1938, and the Industrial Development Fund, set up in 1944. All three agencies provided credit, purchase of private corporate ownership shares, and state financial guarantees to investors; in addition, the CVF served as an investment and a fiduciary manager. The CVF was particularly important, as it received more than 10 percent of the federal budget between 1946 and 1948. It also was closely tied to the national bourgeoisie, and its five-member board of directors consisted of one presidential appointee (Castillo), who was a pro-AD businessman, two congressional appointees, and two nominees selected by Fedecámaras.

In manufacturing, the shoe, cement (Mendoza), and fertilizer industries all received substantial aid; in agriculture, the emphasis was on the development of agribusiness in milk, sugar, and meat. Finally, the governments of this period established a rudimentary financial infrastructure, as a central bank, the Central Bank of Venezuela (BCV), and a stock exchange came into being.

Ambitious as the trienio government's oil policies were, there were certain structural constraints on how much could be accomplished at the time. Because Venezuela's industrial base and home market were still so undeveloped, the nation remained extremely dependent on the oil export sector. The leadership of AD favored incremental gains in capturing the share of oil profits accruing to the state rather than outright expropriation. In the absence of a substantial nonoil sector, Rómulo Betancourt himself advised that Venezuela would make no "suicidal leap into space."[30]

Although many of the AD policies in this period were farsighted and advantageous to the emerging industrial bourgeoisie, the government had insufficient personal-institutional ties with the commercial and industrial ruling class and lacked that class's acceptance of its petty bourgeois democratic ideology and its political methods of mass mobilization. Portions of the rural landed class that were ideologically very conservative were particularly disturbed by the party's anticlerical attitude and its plans for the redistribution of rural property, and some were suspicious of Betancourt's previous flirtations with the Communists when he was exiled during the last years of the Gómez dictatorship. Foreign oil interests were antagonized by AD's oil policies and some industrialists were dissatisfied with their increasing labor costs due to the proliferation of the AD unions. The adecos earned the enmity of the Christian Democratic (COPEI) and the Democratic Republican Union (URD) parties due to its sectarianism, its party patronage in appointments, and the conviction of 98 former government officials for corruption (the more prominent of whom were López Contreras, Medina Angarita, Uslar Pietri, and Pedro Tinoco, Sr.). There were also elements in the military that were hostile to the AD party. In addition to the purged Andean leaders, some resented the party's attempts to politicize the rank and file soldiers and saw it as an attempt to undermine the autonomy of the armed forces. Many others feared the party's armed militia as an alternative military force. When the civilian opposition, particularly COPEI, approached its conservative allies in the military, it found them receptive to the idea of a coup, and, in 1948, the Gallegos government came to an abrupt end.

THE ALIENATED STATE

Capitalist Development in the 1950s

Irrespective of coups d'etat, Venezuela became increasingly integrated into the production, realization, and appropriation of surplus value on a world scale. The Korean War followed by the Suez Crisis in the mid-1950s stimulated a sustained increase in the demand for Venezuelan crude that caused the oil revenues of the Pérez Jiménez government to double between 1951 and 1956.[31] The nation's class structure and large state bureaucracy determined that much of the increased national wealth would accrue to the urban salaried professionals. Given this class's high propensity to consume, there resulted an increased demand for expensive consumer imports.

Due to the accelerated pace of technical innovation brought about by World War II, by the 1950s, U.S. capitalists were plagued with a problem of accumulated surplus capital in the form of obsolete machinery. Venezuela's burgeoning demand for consumer goods, coupled with the social stability, cheap labor, and an open door to foreign capital ensured by the Pérez Jiménez dictatorship, presented an attractive opportunity for U.S. investors. The importation of U.S. capital goods by Venezuelan branches of North American multinationals, which were set up in order to produce consumer goods for the internal market, was the substance of much of Venezuela's import substitution originating in the 1950s.

In global terms, the home market in consumer goods grew appreciably during the period of the Pérez Jiménez regime, as the following table[32] shows.*

Local Supply as the Proportion of Consumer Demand

	1950	1953
	(in percent)	
Foodstuffs (includes processed)	64	74
Other nondurable consumer goods	47	72
Durable consumer goods	12	25
All consumer goods (excluding food)	40	63
All consumer goods (including food)	52	68

Source: Memória, Banco Central de Venezuela, 1959

There was no decrease in the level of imports in this period, only a change in their composition. Primary materials and intermediate products continued to represent 40 percent of total imports, but the proportion of nonoil related capital goods increased significantly at the expense of nondurable consumer products.[33] Although the dependency of the Venezuelan economy on foreign imports was by no means reduced, the extension of the home market was significant in that it led to the further articulation of national social classes corresponding to the capitalist mode of production.

Between the death of one caudillo in 1935 and the political demise of another in 1958, the Venezuelan gross national product grew at an overall rate of 8 percent per year, with agriculture trailing at 4 percent. Manufacturing output grew at 11 percent per year, although unevenly among industries, and, by 1958, had increased its share of the gross national product to 14.6 percent.[34] Between 1948 and 1953, oil refining quadrupled and for the entire period of

---

*The figures in the table may be exaggerated; according to data collected by the Center for Development Studies (CENDES) at the Venezuelan Central University in Caracas, the percentage of Venezuelan food consumption from imports was 10 percent in 1937, 45 percent in 1958, and 24 percent by 1973.

1948 to 1958, growth was substantial in metal fabrication, synthetic rubber, chemicals, plywood, paper, paints, auto assembly and parts, fertilizer, and cement.[35] The tremendous boom in electric energy and construction[36] was the stimulus for the expansion of an intermediate industrial goods sector in building materials and electrical installation and repair. The growth of manufacturing and the increasing number of urban employees created a demand for industrial raw materials and processed food that expanded at 10 percent per year; of particular importance in this sector were wood, sugar, meat, and milk.

The lines demarcating the sectors of economic activity of the Pérez Jiménez government, national capitalists, and foreign investors are quite clear. The government was the principal investor in fixed assets, contributing 40 percent of the total (60 percent, if the oil sector is excluded from the calculation).[37] Basic industry (iron and steel, petrochemicals, electricity) and the infrastructure (public housing, transportation, land colonization) received most of the public funds. Over two thirds of the investment by private national capital went into commerce, housing, and services;[38] the remainder was divided among manufacturing, electricity, primary goods, and commercial agriculture.

By 1958, foreign capital comprised 15 percent of total industrial investment. More important than the aggregate amount was the concentration of such capital in strategic areas of the economy and the share of total surplus value it appropriated. The most profitable industries and the only ones producing significant amounts of foreign exchange—petroleum extraction and iron mining—were 100 percent foreign owned, and nearly entirely by U.S. capital. About one fifth of the major manufacturing areas, especially in such important industries as auto assembly, chemicals, and commerce, were foreign owned. Even the bailiwick of the national capitalists, construction, was not immune from this trend; only 50 percent of such investment was from indigenous capitalists in the 1950s.[39] In addition, following the right-wing coup of 1948, foreign capital was invited into new areas of the Venezuelan economy. The Rockefeller-controlled International Basic Economy Corporation, for example, was given concessions in agriculture, cattle raising, and fishing.

In the countryside the penetration of urban capital supported by government credits began, in the 1950s, to transform commercial agriculture for the first time. Investments by Maracaibo banks and import merchants into the Zulia region were especially important. Extensive, large-scale livestock grazing and capital-intensive mechanized agriculture required large units of land to produce efficiently. An acceleration of the expropriation of the small holder took place; a few remained as wage laborers, many were reduced to a subsistence level of farming, and others were pushed inexorably toward the cities—hoping to sell their labor power.

During the 1940s, there had remained numerous latifundista who utilized their own wealth to invest in agriculture (and often in urban commerce and real estate as well) and who relied on rather traditional production relations. In the Pérez Jiménez period, urban industry and agriculture became much more closely tied, with industrialists-bankers assuming a dominant position. The most direct links were, and have been, those of ownership, of which the Vollmer-Zuloaga interests is an instance. As mentioned earlier, the group's own sugarcane plantations provided the raw material for its refineries. Outside of agriculture, another case of such vertical corporate integration is that of the Mendoza group's Venepal enterprise, which owned its own forests for supplying its paper factories with lumber. In such cases the distribution of surplus value among the different branches of production is decided outside the market by the group's managers as an internal corporate decision.

Many commercial farmers, typically those producing tobacco, cotton, and milk, have been tied to industrial buyers via their dependence on credit. In the 1940s, the commercial farmers commonly relied on such advances. Since then, although there have been larger absolute levels of government credit to agriculture, an increasingly large share of farm credit has been coming from private banks. For instance, the Agricultural Development Bank has taken over some functions of the government Agriculture and Livestock Bank in the sale of agricultural credit. Here the ties of control are quite direct, although the exercise of the privileges of ownership is often mediated as the banks themselves form parts of the various finance capital groups.

An example of a somewhat less direct link between industrial and agricultural capital is provided by the Mendoza finance group. In addition to being a food-processing monopoly, its corporation, Protinal, manufactures and sells farm implements. The group can strongly influence the home market terms of trade between the cities and the countryside by its exercise of market power in both buying from and selling to independent farmers and ranchers.

The relations of production changed as much as the sectoral structure of the economy in the 1950s. Factory production, while still not employing anywhere near a majority of the industrial work force, continued to expropriate the labor-intensive artisan shops. Between 1953 and 1962, the number of industrial establishments decreased from 16,452 to 7,531.[40] An indication of the increasing productivity and economic power of capitalist industrial production was the increases in average total capital per firm: from Bs. 116,000 in 1953 to Bs.1,320,000 by 1966[41]—a more than tenfold jump. While the large factories wiped out their smaller competitors, the small workshop was not eliminated as a form of production; rather, the nature of the articles produced changed from such products as woodwork, clothing, and food to semielaborated goods, for example, spare parts for automobiles. As a result, such workshops were turned into virtual economic vassals of capital, dependent on the industrial monopolies for business, price levels, and credit.

The process of proletarianization increased as more workers labored for fewer firms. The average number of workers per firm in 1936 had been 6, in 1953 it was 9, in 1961 it became 21, and by 1966 was 27. However, the increment in the organic composition of capital (ratio of constant to total capital) from 2.67 to 4.11 between 1953 and 1966[42] suggest why the productive apparatus was unable to absorb most of the rural migrants who were flocking to the cities in hopes of selling their labor power.

While capitalist production in manufacturing had become dominant by 1958, in agriculture this mode controlled large areas of production even where the relations of production between owners and workers on the particular farms and ranches had not as yet been completely transformed to the free labor form. In commercial banking the 1950s was the decade of greatest growth, as deposits increased along with the expanding amounts of oil revenue that were accruing to the Pérez Jiménez government. Between 1950 and 1959, the total credit and money supply together multiplied three and one-half times, while commercial bank credit enjoyed a sixfold increment.[43] In spite of such additions to the supply of liquid capital, bank and industrial monopolies were prevented from coalescing completely into the finance capital form due to antiquated bank laws, which would be removed only after the 1958 coup.

The Constraints on the Accumulation of National Capital

The 1948 coup represented a step backward for the attempts of Venezuelans to institute capitalist democracy. The new government, led by Colonel Pérez Jiménez and Lt. Colonel Delgado Chalbaud, began immediately to implement repressive political measures. The 1936 constitution, with its retrograde limits on the electoral franchise, was reinstated and the AD was declared illegal and its leaders exiled. The style of government was changed from one of mass mobilization to one of personalism and nepotism. The FCV was disbanded and the CTV replaced by the government puppet confederation, the National Confederation of Workers (CNT); labor organizing, strikes, and collective bargaining were sharply curtailed. The new government's policies constituted a direct attack on the gains made during the trienio period by the working classes: land reform was halted, the secular education policy was reversed, and the cost of living in basic necessities was allowed to rise.

In the two years following the coup, Pérez Jiménez provided the military leadership and Chalbaud the ties with the propertied classes. By 1950, Chalbaud had been assassinated and Pérez Jiménez and the army were in complete control. After 1950, most of Pérez Jiménez' limited support came from the Catholic Church hierarchy (initially), the unorganized subproletariat of Caracas, and from real estate speculators and construction contractors who profited from the government's many public works projects in the capital. Even with the narrow property franchise, the government party, the Independent Electoral Front (FEI), was defeated in the 1952 elections by the URD in coalition with COPEI and the outlawed AD, and the government was forced to nullify the results. The only other attempt at legitimizing the dictatorship occurred in

1957 in a sham plebiscite that only served to underscore the regime's estrangement from the population.

With respect to the oil sector, the 1948 coup ended the collection of royalties in crude oil to be marketed by the then-existing CVP. The government oil corporation was itself eliminated, the anticyclical public fund discontinued, the government's taxation of oil revenues reduced, and new concessions opened up. Also, those ministers of ex-President Medina Angarita who had had connections with the foreign oil companies were placed in the new cabinet and there were assurances made by the military regime that there would be no change in the oil law. The establishment of such a congenial investment climate cannot be attributed solely to the nature of the military government. There was a good deal of coercive economic leverage applied by the oil corporations: the level of oil production was reduced by 15 percent following the coup and Chase Manhattan Bank threatened to withdraw capital. The government and the economy were reduced to total dependence on production levels, concession sales, and the price of crude oil. Given the unwillingness or inability of the government to collect public revenue through either income or corporate taxes, such dependency put Pérez Jiménez in a very precarious fiscal position in the late 1950s after the oil boom ended.

In regard to the nonoil sectors of the national economy, Pérez Jiménez undertook a modest import-substitution program. Selective ad valorem tariffs were used to protect infant industries in consumer nondurables, intermediate goods, and those foodstuffs and raw materials that were processed by existing or protected national industries.[44] In the previous era of free trade imperialism, such import substitution would have constituted a defense of the interests of a national bourgeoisie. However, given the levels of direct foreign investment in Venezuela following World War II, such a limited import-substitution policy had less importance. Of greater relevance for Venezuela since the 1940s was the policy allowing for the free convertibility of bolivars into foreign currency. It was such conversion that facilitated the repatriation of the profits and service payments of the Venezuelan branches and subsidiaries of multinational corporations. It was this policy, combined with a very high level of liquidity, which was to lead to the massive flight of capital from Venezuela in 1958 and 1959.[45]

Pérez Jiménez' program for developing the national economy was centered around three major areas. *Commercial agriculture* was supported by government subsidies and credits to farmers and import protection. Development of the *infrastructure* was emphasized and priority was placed on transport services and public buildings. Rails were deemphasized, but roads, ports, and airports were built. Much of the construction, however, was in wasteful, prestige projects, such as the Caracas race track, 12 luxury hotels, and plush officers' clubs. *Basic heavy industries* in, for example, electricity (the Company for the Administration and Supply of Electricity [CADAFE]), petrochemicals (the Venezuelan Petrochemical Institute [IVP]), and steel were established and expanded, even if not always with notable success.

The key policy failure of the Pérez Jiménez regime from the viewpoint of the national bourgeoisie was the insufficiency of government oil revenues made available to the private sector for long-term industrial investment. The primary sources of private capital (aside from retained earnings) were the commercial banks,* the BIV, and the CVF. The commercial banks were actually prohibited by law from making loans of more than two and one-half years duration or from discounting commercial notes with more than a year's maturity.[46] Such a highly liquid supply of capital could benefit only Pérez Jiménez' main supporters: the commercial and real estate speculators.

The CVF under Pérez Jiménez served to channel the oil-derived economic surplus to the state enterprises and to restrict importation by limiting the number of import licenses granted by the government. The overall budgetary allocations were similar for the Pérez Jiménez and the provisional governments and the ratio of total investments to the gross national product was about the same. However, there was a crucial difference in the method of distributing the capital expenditures of the CVF, as the following table shows.[47]

Distribution of Capital Expenditures

Year	Direct (Percent)	Indirect (Percent)	Total Amount (million Bs.)
1952-53	26.56	11.47	90,548
1953-54	29.71	8.38	94,301
1954-55	37.41	7.41	128,886
1955-56	38.46	6.68	137,648
1956-57	47.42	6.31	210,632
1957-58	54.37	4.78	360,774
1958-59	30.04	20.14	332,118
1959-60	23.58	16.29	267,940
1960-61	16.62	16.21	233,616
1961 (June-Dec.)	20.32	18.97	125,208
1962	16.90	12.80	1,944
1963	18.80	10.20	1,924

Pérez Jiménez placed a preponderant and increasing emphasis on direct aid to state enterprises, while the provisional government of 1958 established a parity between this method and that of indirect investment financing of pri-

---

*Private investment banks did not exist before 1958.

vate capitalists, that is, increasing special government funds deposited in private investment banks and by outright purchase of stock in commercial banks.

Whereas the 1948 coup had been the result primarily of antagonisms within the ruling class, the overthrow of Pérez Jiménez in 1958 was caused by the contradiction between an emerging industrial bourgeoisie and the existing Venezuelan state. Pérez Jiménez' corruption and incompetence could have been endured, but members of this class feared their displacement by the increasing direct foreign investment and they correctly perceived the government as being ill-equipped to handle, and unresponsive to, their needs as a national ruling class. The coup was precipitated by the end of the oil export boom. Given the dependency of capital accumulation on the level of oil exports, which Pérez Jiménez had allowed to develop, the inadequacy of the government became manifest as economic growth slowed and the reliance on foreign financing became more onerous. The further development of national capitalism required the Venezuelan bourgeoisie to create a specifically capitalist state and to ensure access to it by members of the class.

The finance capitalists, especially Mendoza and Vollmer, who helped initiate the polyclass movement against the military government, suffered concrete losses at the hands of that regime. In the early 1950s, for example, Mendoza and Vollmer initiated plans for the building of a small steel plant. At the time of the completion of the plans Pérez Jiménez forbade the establishment of the proposed firm, informing the capitalists that the state was planning a larger enterprise. The government then set up a contract with an Italian firm and began producing steel tubing, for which there was to be scant demand due to declining levels of oil exploration.

Although there were elements in the military dissatisfied with the corruption and general mismanagement of the Pérez Jiménez regime, the 1958 coup was largely civilian-initiated and organized. It represented a broad political alliance that included liberals within the Catholic Church, a university front, the major political parties, and those urban leftists who on the eve of the overthrow organized a successful general strike. The leadership of the Patriotic Junta, however, came from among the monopoly bourgeoisie. In the year following the government's fiasco over the initiation of a Venezuelan steel industry, Mendoza and other leading national capitalists began supporting a coup against the government. About the same time, Vollmer and Mendoza assumed leadership roles within Fedecámaras, which had been having difficulties arriving at consensus. It was Fedecámaras that served as the liaison between big business and the AD politicians during the political transformations between 1957 and 1960.

The composition of the five-member provisional government amply demonstrated the predominance of the bourgeoisie in the reconstitution of a different kind of state apparatus.[48] Oscar Centenos Sucinchi was a pharmacist and

Fabricio Ojeda a leftist leader. However, the president of the junta, Requeño, was a representative of the professional guilds (*gremios*), Blas Lamberti was a corporate engineer, and Eugenio Mendoza represented the interests of the finance capitalist groups. Mendoza had been offered the presidency of a proposed post junta counter coup by a far-right general, Leon Castro, but had had to refuse after businessmen of the Republican Integration and Fedecámaras, the United National Labor Committee, and the leaders of the URD (Jóvito Villalba) and the COPEI (Rafael Caldera) refused to support him.

One can infer the primary motivations of the Venezuelan bourgeoisie for making the coup by examining the major policy reforms and changes in the structure of the state made by the 1958 provisional government. The electricity industry was one of the first areas to be significantly affected by the change of government. Shortly after the coup, the public utility CADAFE increased its share of the generation of electric power within Venezuela from 30 to 40 percent.[49] The change benefited private national capital in two ways. First, although its share of total power generation had fallen, with the aid of government subsidies it improved its percentage relative to foreign capital.[50] Second, because CADAFE generally assumed responsibility for supplying energy in geographic areas where its use was the least concentrated, private capital was able to focus on the more urbanized and industrialized areas and thereby greatly increase its rate of profit.[51]

With the nation's economic infrastructure developed beyond the capacity of the national industrialists to utilize it, sections of the Venezuelan bourgeoisie (such as the businessmen's Pro-Venezuela Association, formed in 1958) agitated for increased import-substitution-based industrialization and more state planning. In the same year, the state planning agency, CORDIPLAN (Central Office of Coordinator and Planning), was established, and by 1960, the AD government had increased substantially the level of trade restrictions.

The major obstacle to capital accumulation, limited access of the capitalist class to long-term investment capital, was removed by the provisional government. The CVF continued as one of the primary development agencies, but there was a change in its industrialization policy. After 1958, it channeled an increasing share (and amount) of its state investment funds to the financing of private, and predominantly national, capitalists.[52] To ensure the bourgeoisie of continued access to this public corporation, ownership shares were made available for purchase by private institutions and individuals, and today it is controlled by the nation's large finance capital groups. Venezuelan capitalists were also able to tap additional amounts of private wealth (after 1958) for their investments due to the changes in the banking laws. Commercial banks were allowed to provide long-term loans and discounted notes. Investment and mortgage banks, under the control of the groups that owned the commercial banks, emerged as major money sources for portfolio investments and for mortgages.

From National Populism to National Capitalism: AD's Return to Power

When AD replaced the provisional government and organized a parliamentary democracy in 1959, it did so with the lessons of 1946 to 1948 well in mind: the party leadership was determined to hold power in addition to winning elections. In any case its electoral base was considerably weaker than it had been during the trienio. The urbanization and proletarianization (or at least expropriation) of much of the nation's work force had at once changed the nature of the key political issues and undermined the party's traditional base of support. Building political programs around such petty bourgeois issues as the extension of the franchise to small holder farmers would no longer be a viable strategy.

Because AD was elected to political power in 1959 as a minority government, it enhanced the stability of the new regime while ingratiating itself with the bourgeoisie by forming a parliamentary coalition with the center-right parties (COPEI and URD). Besides excluding the Venezuelan Communist party from participation in the ruling coalition, AD signaled that it envisaged no radical changes in the economic structure. Furthermore, once in office it strongly undercut any nationalist posture by assuming large debts to the United States in order to extricate the government from the precarious financial position it had inherited from Pérez Jiménez. As the party with the strongest organization and the most clearly articulated program, the AD was preferred by most members of the nation's ruling class to provide them with their initial entry into the state apparatus in order to make the changes in its structure that they desired.

The AD platform of 1959, the *Bases Programaticas*,[53] expressed the party's promise as a vehicle for a more effective bourgeois rule. The *Tesis Agraria* plank stressed the intent of the party to modernize and increase the productivity of agriculture. Rural credit was to be provided to efficient farmers, while the foot-dragging latifundista would be in danger of expropriation, but only where compensation was provided. The landless were to be colonized on small parcels of government land and organized into cooperatives. The *Tesis Sindical* (rhetoric about eventual democratic socialism aside) posited that the trade unions, allied with parliamentary political parties, were to be the designated vehicle for petitioning the ruling class for limited and economistic improvements for wage earners. In its foreign policy the party was not antiforeign investment, but through the use of tactical alliances with organized labor it wished to rely on state regulation and direction of foreign capital. Specifically, it advocated a decrease in the nation's dependence on oil and iron exports, claimed national subsoil rights, and suggested the establishment of domestically owned oil refineries. The *Tesis Petrolera* proposed the re-creation of the CVP, but deemed nationalization inappropriate.

The 1959–62 period of the first civilian government was one of double digit unemployment and a generally severe recession. Pérez Jiménez also had had to cope with the problem of idle, hungry migrants in Caracas; his stopgap solution had been to provide public works employment in extravagant contruction projects. This tactic ameliorated conditions of unemployment as long as the government oil revenues held out, but declining Venezuelan oil production during the Betancourt government made this policy unfeasible as long as AD was not willing to alienate the military or the bourgeoisie by reducing budget allocations crucial to its interests. As a result, AD got less than 15 percent of the Caracas vote in the electoral contests in both 1958 and 1963.

The party continued moving to the right and losing popular electoral support during the 1960s* due to the defection on three separate occasions of the more progressive elements within the party. The first took place in 1960 and was precipitated by radicalized segments of AD's student supporters, who left the organization in opposition to Betancourt's loans from the International Monetary Fund, the World Bank, and a New York City bank consortium, and as a result of the government's decision to sever diplomatic relations with Cuba. The 1960 dissidents became the core of the Movement of the Revolutionary Left (MIR). The second split occurred in 1962 and was brought about by the more liberal leadership within the FCV and the CTV. This group formed its version of the party as AD opposition, later reconstituted as the Revolutionary Party of Nationalist Integration (PRIN). The third schism was in 1967 over the selection of AD's candidate in the 1968 presidential election. This last defecting group, the People's Electoral Movement (MEP), was the only one of the three to have a significant electoral impact on AD.†

Pérez Alfonzo, who, first as legal counsel to the provisional government in 1945 and then as minister of development during the trienio, had been the initiator of many of the previous reforms of the petroleum industry and became even more influential in the AD and COPEI administrations regarding the nation's oil policy. He became minister of hydrocarbons and mines in 1959, and stressed the conservation of oil deposits, controlled levels of production, higher taxation (in 1958 he raised the rate to 65 percent), higher estimates on the base price used for the calculation of the tax, and public regulation of the multinational oil corporations at both the national and international levels (Pérez Alfonzo was the impetus behind the formation of the CVP and OPEC

---

*AD's presidential candidates in 1963 and 1968 (Raúl Leoni and Gonzalo Barrios, respectively received 33 percent and then 28 percent of the vote, with Barrios losing to the COPEI candidate, Rafael Caldera.

†MEP polled 13 percent of the congressional vote in 1968 and its presidential candidate, Luis Beltran Prieto, received 19 percent, taking enough of AD's support to allow Caldera of COPEI to win the election.

in 1960). Even after he was replaced in office after the COPEI electoral victory in 1969, he maintained personal ties with the legislature and continued to be influential in the drafting of subsequent oil legislation.

THE CONSOLIDATION OF STATE CAPITALISM

Diversification of the Enclave Economy: The Limits of Import Substitution

Due to the continuing reliance on the state oil revenues for the financing of the nation's industrialization, the rate of growth of the Venezuelan economy remains today largely a function of the world demand for petroleum. During the oil boom of 1950–57, Venezuela's gross national product grew at a rate of 12 percent per year. From 1957 until 1970, the nation's aggregate production increased, on the average, 7 percent yearly. The figure for this period was reduced largely as a result of the Venezuelan 1959–62 recession brought on by the substitution of Middle East for high-sulfur Venezuelan crude by the petroleum-importing countries.[54] Between 1970 and 1973, another period of declining oil prices, growth dropped to 4.2 percent per year,[55] while, since 1973, national production has increased considerably in response to the more than sevenfold jump in the price of a barrel of oil.[56]

In addition to the aggregate growth that has continued since the overthrow of Pérez Jiménez, the structure of production has changed significantly. By 1973, the profile of the nation's gross national product was somewhat different than it had been in 1950:[57]

Service	50%
Petroleum	21
Agriculture	6
Construction	5
Mining	1
Electricity	1
Manufacturing	19

The contribution of agriculture had decreased, petroleum and construction suffered slight declines, slight gains were made by mining and electricity, and manufacturing's share increased a considerable 9 percent. Private capital flowed out of agriculture, transportation, electricity, and petroleum into manufacturing; overall, the primary sector increased its output 50 percent by 1970, while the secondary sectors doubled and the tertiary sector expanded even more.

The impetus for the growth of the home market in the post-Pérez Jiménez period had come from intensified import substitution and the growing volume of consumer credit. Because of the considerable foreign exchange generated by the oil exports, Venezuela has continually enjoyed a substantial import capacity, unlike most neocolonial nations that have experienced the penetration of foreign capital. Following the change in 1958, Venezuela's level of imports rose and, in response to the successive governments' attempts to foster industrialization, their composition continued to shift away from consumer goods.[58] Unlike the experience of other Latin American countries, Venezuela's import substitution occurred primarily in the 1960s instead of the 1930s and 1940s. The ultimate limits of the process, the size of the consumer market within the nation, were reached relatively much more rapidly. This occurred primarily because the economic infrastructure already existed, due to the abundant state financing of private investment and because the foreign multinationals were able to mobilize quickly large capital resources in order to exploit this golden opportunity to produce within the protected Venezuelan market. The easy stage of industrialization—production of consumer goods for the upper middle income stratum—which offers quick returns on invested capital and is less technologically demanding, was accomplished by 1970. Nonetheless, judged solely by the criterion of import substitution, expansion of an industrial home market, the program accomplished its goal. Whereas, in 1960, 65 percent of the domestic consumption of industrial goods had been supplied locally, by 1970, the share was 85 percent.

In spite of the Keynesian fiscal and monetary policies pursued by the succession of governments after 1958, Venezuelan and foreign investors soon encountered problems of overproduction in the attempts to industrialize. High rates of capital formation and exploitation combined with the labor-saving technology used by the large domestic and foreign investors and the income concentrating policies of the capitalist state all have exacerbated the problem of narrow consumer goods markets. The results have been excess industrial capacity[59] combined with overall low productivity and high costs, the separation of growth and high productivity sectors from those providing the bulk of national employment,[60] increased penetration of foreign capital leading to foreign appropriation of the surplus value produced by the Venezuelan proletariat[61] and a loss of national control over the Venezuelan economy, the diversification of industrial production primarily in terms of the multiplication of brands and models of luxury consumer goods, the expansion of business services and the stagnation of social services, and chronic unemployment and underemployment.[62]

Rather than attempt to redistribute income and create employment, the Venezuelan governments have tried (unsuccessfully) to respond to the drying up of further import-substitution possibilities by promoting the diversification of exports and by trying to rationalize and integrate production for the internal market. Private capital has attempted to alleviate the overproduction problem by expanding consumer credit in durable goods.[63] For instance, increasing production by the foreign-owned auto assembly industry has been absorbed by purchases financed through finance companies, such as GM Acceptance and FinalVen,[64] and residential construction has been stimulated by the sale of home mortgages by the mortgage banks.[65]

One of the most striking aspects of the Venezuelan economic growth of the 1960s and 1970s was the continuing penetration of foreign capital. By 1971, Venezuela already had the largest gross accumulated foreign investment of any Third-World country: $5.57 billion, of which $5.3 billion was in direct investment and $220 million in portfolio investment.[66] Of the total foreign capital, 86 percent was allocated to the petroleum industry; manufacturing and mining each received another 5 percent and 4 percent was spread among other areas, such as utilities, commerce, and so on. Until the restrictions on foreign investment laid down by the Cartegena Agreement (Andean Pact) in 1968 were later applied by the Caldera government, foreign capital had had largely free access to the Venezuelan economy. In addition to the oil enclave, foreign corporations were important forces in electrical machinery assembly, metal products fabrication, food processing, textiles, glassware, intermediary materials (paper, plastics, chemical products, nonmetal minerals, pharmaceuticals and petrochemicals), construction, commerce, banking-financial institutions, and services (hotels, airlines, advertising). In all, foreign capital was responsible in 1971 for 13.8 percent of all the investment by monopoly firms (100 or more employees) and accounted for 19 percent of total manufacturing profits.[67] It has been the fastest growing areas of manufacturing in which foreign multinational corporations have concentrated their penetration of the Venezuelan economy, thereby preempting national capital.

Among foreign corporations in Venezuela, the U.S. multinationals clearly predominate. In 1970, 71 percent of all foreign investment in the petroleum industry was North American and 22 percent British.[68] In other strategic areas of the Venezuelan economy, U.S. capital represented by 1967 the lion's share of foreign investment:[69]

Mining	100%
Manufacturing	68
Commerce	68
Banking	73
Construction	43

One of the major corporate organizations used to recycle oil profits into the nonoil sector has been the Creole Investment Corporation, a subsidiary of Creole Petroleum Corporation, formed in 1961 to undertake joint ventures with Venezuelan capital. The oil corporations' interest in diversifying their investments in Venezuela is understandable, not only in terms of responding to the increasingly aggressive policies regarding the oil industry pursued by the Venezuelan governments but also in light of the relative attractiveness of profit-taking opportunities that exist outside the oil enclave.[70] The diversification trend has further strengthened the business ties between U.S. and Venezuelan national finance capital: today some 50 percent of all large firms in Venezuela have some financial connection with U.S. private capital.[71]

The political and ideological ties between North American and Venezuelan businessmen have also been cultivated since the 1957 coup. The foreign oil corporations constituted themselves as the Chamber of the Petroleum Industry and were allowed representation in Fedecámaras in 1959. Since then, the oil corporations, in addition to providing technical advice in the formulation of the Fedecámaras' oil policy, have been that organization's largest financial contributor and today two of the twenty seats of the Fedecámaras' board of directors are held by these foreign corporations. Also, the Creole Petroleum Corporation established the Creole Foundation, which, in 1964, combined with the Venezuelan Association of Executives and the Mendoza Foundation to form Voluntary Dividends for the Community (DIVIDENDO). DIVIDENDO has served to legitimize the domination of finance capital in Venezuela by contributing money for educational and social welfare projects.

If, as is argued, the Venezuelan state is strongly capitalist in both its structure and in the degree to which it is open to the influence of national business interests, and if the considerable penetration of both the oil and nonoil sectors of the Venezuelan economy by foreign capital has been so extensive as to create strong ties between the national monopoly bourgeoisie and foreign investors, then the question arises as to the structural conditions of Venezuelan capitalism that allowed for the possibility (though not the necessity) of the oil nationalization.

The first condition was the oil sector's constitution as an economic enclave. Despite the influx of foreign investments into the nonoil sector of the Venezuelan economy, national capitalists had few direct ties to investments in oil. Their relation to the oil sector had been largely via the state, which they used to extract revenue from the foreign oil corporations in order to finance their own investments in the nonoil sector. For the most part, the oil sector existed as a set of self-contained and internally integrated economic activities. Foreign investment in petroleum totally dominated every aspect of production and marketing. The investments of the foreign oil companies were diffused throughout the sector by 1973:[72]

Extraction of crude oil	70%
Refining	10
Transport	12
Marketing	8

The private Venezuelan oil company, Mito-Juan, has never had a significant volume of production, and the direct ties of the Venezuelan industrial bourgeoisie to the oil sector via its purchases of oil for its factories were further limited in 1967 when the public CVP took over one third of all oil sales within the domestic market.

The second factor was the growth of production of the nonoil sector, both in absolute terms and relative to that of the oil enclave. U.S. and British corporations continued to sink capital into Venezuelan petroleum in the years following the Pérez Jiménez regime. However, such new investment, primarily in refining and transport, did not keep pace with the depreciation of existing capital and the drawing down of the reserve stocks of crude oil. The result has been declining levels of net investment in fixed assets in a period in which the world volume of petroleum investment was increasing:[73]

Net Investment in Fixed Assets (millions of bolivars)

	Production	Transport	Refining	Sales	Other	Total
1957	6,681	721	1,181	102	317	9,002
1958	6,966	993	1,215	96	382	9,652
1959	7,534	1,067	1,199	107	468	10,375
1960	7,095	1,032	1,108	117	420	9,772
1961	6,658	1,074	1,002	117	415	9,266
1962	6,203	1,047	909	120	385	8,664
1963	5,785	997	810	117	377	8,086
1964	5,545	1,011	766	127	329	7,778
1965	5,486	954	675	121	317	7,553
1966	5,065	918	612	120	328	7,043
1967	4,721	864	573	124	294	6,756
1968	4,928	856	646	118	305	6,853
1969	4,966	946	934	116	399	7,361
1970	4,790	916	1,342	115	298	7,461
1971	4,651	906	1,328	114	293	7,292

The annual increases in national production[74] reflected the growth of the nonoil sector as foreign investors began to look with a covetous eye on the nation's increasing political stability, growing internal market, and the Venezuelan government's continued channeling of public oil revenues to domestic investors:

	1950–57 (in percent)	1957–68 (in percent)
Value of nonoil production	9.26	5.32
Value of oil production (1957 prices)	9.65	1.99

The enclave and growth factors, when combined with the diversification of the investments of the foreign oil capitalists into the nonoil sector, made nationalization an increasingly efficacious policy in both economic and political terms. Because the vital interests of the national monopoly bourgeoisie were not tied to the oil sector, this class would oppose nationalization in principle (for fear that such actions might arouse working class demands for similar actions toward national capital), but would make no militant obstruction of the process as long as it remained sectoral. After the petty bourgeois nationalists created the political conditions for nationalization, the monopoly bourgeoisie moved to appropriate the surplus and integrate itself with foreign capital on new and more lucrative bases, for example, partnerships in the manufacture of new petrochemical products for exportation.

During the trienio period of the 1940s, AD, under the leadership of Betancourt, had considered oil nationalization to be a "leap into space." This was due to a minuscule home market, insufficient productive forces, and the lack of a well-developed and politically organized national bourgeoisie wedded to a capitalist state. By 1976, the "space" had been filled.

Impact of Capitalist Development on the Class Structure

Contemporary Venezuela has a clearly defined class structure based on the capitalist wage-property dichotomy. Eighty percent of the population experiences its production relations in an urban context and it is in the cities where the alienation of productive property has proceeded the farthest:[75]

Occupational Categories	Composition of Work Force (1974)
	(in percent)
Agricultural activity	21.1
Employee-worker	8.3
Employer	1.1
Self-employed	7.5
Family helper	3.5
Nonagricultural activity	78.8
Employee-worker	57.9
Employer	3.2
Self-employed	15.9
Family helper	1.8
Total	100

The urban productive forces have demonstrated a singular inability to absorb the expanding work force into productive activity. From 1950 to 1973, while real gross national product increased four and one-half times and the work force doubled, services employment expanded two and one-half times—growing from 34 to 52 percent of the total work force. Under this general rubric, two categories in particular grew rapidly: small merchants and government employees. The number of street venders tripled in the 23-year period—reaching 18 percent of total employment, while government employees represented the same share—up from 8 percent as late as 1961. Presently about 17 percent of the economically active population is employed in manufacturing and over one third of these Venezuelans labor in small and often low productivity firms of 5 to 20 workers.[76]

In the countryside, the industrialization of the last two decades has stimulated the commercialization of agriculture and in doing so has prevented the establishment of a more egalitarian distribution of land. Four major rural classes have been formed in the process: *Agricultural workers,* those who rely predominantly on the sale of their labor power to secure their means of subsistence, by 1974, constituted nearly 40 percent of the work force engaged in agricultural activity. Given that there has been minimal out-migration from the cities, it appears that a substantial number of Venezuelans remain, or have been, alienated from the land. *Small holders* are those who work their own land and hire either very occasional wage labor or none at all; they represent 80 percent of all farm owners and toil on 5 percent of the cultivated land. *Middle farmers* work the land but also hire labor on a fairly regular basis; they comprise 15 percent of all farm owners and hold 15 percent of the cultivated land. *Landlords* rely on the labor of others, dominate local political and administrative organizations, and monopolize both government and private sources of agricultural credit; they are 5 percent of all farmers and their properties constitute 80 percent of the cultivated land.[77] In terms of national farm production in 1973, 80 percent came from the middle to large landowners and 12 percent from the small holders; the agrarian reform landholders (numbering about 100,000 in the same year) were responsible for the remaining 8 percent.[78]

The belated but rapid development of capitalism in Venezuela has resulted in changes of the class distribution of the national product. This has been manifested in the relative shares of national personal income going to property (profits, interest, rent) and labor (wages, salaries). In 1950, the division was property income 40 percent and labor income 60 percent. By 1973, property accounted for 53 percent and labor for 47 percent of national personal income.[79] Of course, this simple dichotomy obscures much about the class distribution of income in Venezuela. A good deal of the wages and salaries allotment undoubtedly went to the salaried petty bourgeoisie: professional and well-remunerated government employees. Similarly, property income includes some of the profits and rent of the self-employed and perhaps the interest accruing to the numerous small rural usurers.

The notion that Venezuela's government oil revenues have created a sizable upper middle income stratum of public employees is buttressed by a comparison of Venezuela's income distribution in 1970 with that of six other Latin American nations.[80] The median reported income of the richest 5 percent of the Venezuelan population ranks the lowest of the seven countries; the average income of the 80th to the 95 percentile ranks the highest; and the nation's bottom 20 percent of the income distribution rates as relatively the worst off among the seven.

It would also appear that the democratic revolution of 1958 has had relatively little effect on Venezuela's income distribution:[81]

National Personal Income Distribution (by families) (in percent)

	Top 5	Middle 45	Lowest 50
1957	25	56	19
1962	24	59	17
1970	22	58	20

On the other hand, within Venezuelan industry in recent years the processes of capital accumulation (concentration and centralization) and of proletarianization have continued. Between 1961 and 1971, industrial productivity increased 72 percent, while the money value of the wage bill (that is, not adjusted for inflation) climbed by 60 percent, which allowed the rate of factory exploitation to go from 158 to 177 percent.[82] Increased exploitation and the uneven distribution among small and large firms of state-financed industrial credit have helped consolidate the position of the industrial monopolies:[83]

Industrial Monopolies' Share	1961	1971
	(in percent)	
Employed industrial workers	37	55
Industrial production	62	72
Fixed industrial capital	—	80
Industrial profits	68	72

Even though capitalist development is, by its very nature, an exploitative way to expand the productive forces, it might be imagined that some of Venezuela's increased national wealth would have trickled down to the working class. With the possible exception of improved literacy, however, this appears not to have been the case. In 1974, 67 percent of all nonagricultural employees received wages insufficient to provide a subsistence family income of Bs.1,000 per month,[84] and 29 percent fell below the official minimum wage of Bs.500 per month. Given Venezuela's occupational and class structure and its income distribution, it is clear that it is the majority of the nation's proletariat, subproletariat, and army of self-employed street vendors who experience their social position as one of poverty. Of the total national population, 70 percent suffer from inadequate diets[85] and within the nation's most developed city, Caracas, 28 percent lack running water,[86] 23 percent live without sewers and septic tanks,[87] and nearly a fourth of the city's inhabitants have no electricity.[88]

Regarding the nation's capitalist class, a general outline of its composition by sectors can be obtained from the profile of the 146 member associations that made up Fedecámaras in 1969: 28 in trade, 50 in manufacturing, 24 in manufacturing and trade, 5 in manufacturing and trade and agriculture, 15 in commercial farming, 9 in commercial ranching, and 15 in services.[89] The Venezuelan bourgeoisie exhibited considerable political unity in the face of events, social forces, proposed government policies, and so on that threatened (or were perceived to threaten) the class as a whole. Two specific instances were the urban guerrillas during the early 1960s and the legislation proposed originally in 1966 to increase personal and corporate income taxation and to institute an excess profits tax.

However, intraclass conflict has occurred frequently over the issues that affected the various sectors in different ways; for example, membership in the Latin American Free Trade Association (LAFTA), proposed in 1967, and entry into the Andean Pact, were issues in 1968. It was also in the 1960s that separate capitalist interest groups began to devise political formations as alternatives to Fedecámaras (which is dominated by the financial oligarchy in alliance with foreign capitalist interests), for example, Pro-Venezuela and the Venezuela Association of Executives.

The direction in which Venezuelan capitalism has developed has caused the formation of a national class structure with discernible political implications. Because of the way in which the Venezuelan population has historically had its economy integrated with world capitalist accumulation, much of the private productive activity within present-day Venezuela is devoted to the tertiary sector. As a result, many initiatives in production (and political initiatives) have had to come from either foreign capital or the state. Also, the strong concentration of national finance capital and its mercantile origins have allowed the establishment of ruling class political institutions (including the state itself) that often rely on direct, informal ties between the bourgeoisie and the agencies of class rule. Furthermore, the existence of a large subproletariat and the sharp divisions between the domestic and foreign-controlled sectors of the economy and between monopoly and competitive firms have created disparate production relations among the working class. The effect has been to divide and weaken the trade unions and peasant league and cause workers to rely on state intervention to provide solutions to their problems. As will be shown, all these political aspects of Venezuela's class structure were manifested in the process of nationalizing oil.

The Changing Nature of the State

The limiting of the scope of the armed forces' activity, although not the intensity (witness the counterinsurgency campaign against the MIR and the Venezuelan Communist party in the early 1960s), was one of the most significant changes in the structure of the Venezuelan state after the rule of the provisional government. In the initial period the urban warfare of the leftist guerrillas actually served the interests of the new civilian government organized by AD: those counter-coup-inclined generals who were still loyal to Pérez Jiménez were forced to devote their energies to suppressing the revolutionaries. Since the guerrillas' defeat, the bourgeois politicians have managed to keep a tight rein on the military. Large military bases, such as Maracay, have been dispersed and the leadership of the armed forces is periodically rotated. The president selects his commanding officers from lists submitted by the different branches of the military and the Congress then ratifies the appointments. The ample public oil revenues have aided in maintaining the acquiescence of the military by providing the civilian governments with the funds for increased professional training, subsidized housing for officers, and so on.

Equally important has been the proliferation of state agencies. The ministerial system, which had begun to be changed in the 1950s, after 1960 gave way to the expansion of autonomous state institutes and enterprises. Whereas, in 1940, there had only existed the Workers' Bank and the Agricultural and Livestock Bank, by 1974, there were more than 80 such public entities—autonomous from the president's cabinet and generally directly facilitating the accumulation of private capital.

The CADAFE, the Venezuelan Guyana Corporation (CVG), the IVP, and the CVP provide the infrastructure and heavy industry that the national bourgeoisie has been unable or unwilling to develop. The National Agrarian Institute (IAN) has had the responsibility for carrying out the land redistribution that was part of the 1960s agrarian reform program. The Office of Integrated Educational Planning (EDUPLAN) has attempted to rationalize and coordinate public education.

A large number of the institutes function to extend the public financing of private investments. The Venezuelan Investment Fund (FIV) is officially described as an autonomous legal entity for the "administrative investment of its own assets for the financing of the expansion and diversification of the economy, making profitable placements abroad and promoting programs of international cooperation."[90] The CVF remains independent of the Ministry of Development; it finances and promotes investments by national or foreign capitalists via direct loans or investments, operates government industrial enterprises through subsidiary corporations, and attempts to rationalize the internal marketing system. The Agricultural and Livestock Bank, the Agricultural Development Bank, and the BIV make direct credit offerings, while the Agricultural Development Fund and the FIV do so through the administration of private commercial banks. By 1970, the autonomous agencies collected 25 percent of total public receipts and accounted for 45 percent of public expenditures while contracting up to 60 percent of the public debt.[91]

Other attempts at capitalist state reforms have not seen fruition, in part because they have been in conflict with the system of autonomous agencies. CORDIPLAN has not been effective; the first four of its plans have been very general and imprecise. Since the legislature is not involved in the elaboration of the plan, it cannot have the force of law and remains merely "indicative." As a result there tends to be little coordination between the plan and the national budget. Also, the president's Advisory Commission of Public Administration tried, unsuccessfully, in 1971, to replace the corrupt, inefficient patronage system with a civil service system based on a 1958–61 study by North American consulting firms.

Class and State

The degree and way in which different classes have access to the capitalist state apparatus have undergone important changes that have had relevance to the formulation of government policies that either by intent or effect (for example, the nationalizations) serve to strengthen the national bourgeoisie. Working class access to the Venezuelan state is minimal. Nominally, its interests are represented by the two major labor union confederations, the Catholic-dominated Committee of Autonomous Unions (CODESA) and the CTV, which are tied by common leadership to COPEI and AD, respectively.

The CTV, the more important of the two, has functioned to keep the unemployed and the proletariat divided, to depoliticize working class demands, and to demobilize workers. For example, in 1958 the Law of Collective Contracts created state-arbitrated collective bargaining procedures on a regional and an industrywide basis, which was to have reduced the likelihood of independent worker initiatives, such as unauthorized strikes, at the level of individual firms. In the context of rapidly increasing unemployment, worker militancy, and the threat of competition from a leftist-oriented labor confederation—the Unified Confederation of Venezuelan Workers (CUTV)—the CTV negotiated a *paz sindical,* or labor peace, with Fedecámaras.

Rural workers have had even less parliamentary influence within the state. The political organization that represents them, the FCV, relies largely on clientelist manipulation of its membership. Not only is the FCV financially dependent on the Ministry of Labor but its directorship is composed of AD party and labor leaders. Neither rural nor urban workers have any input into CORDIPLAN due to the efforts of governments to depoliticize economic planning.

Since the days of the provisional government in 1958, the Venezuelan bourgeoisie has been most successful in gaining direct access to the state via its reach into the agencies that distribute government credit and establish fiscal and monetary policy: key cabinet ministries, certain of the state enterprises and holding companies, and other autonomous financial institutes. The Ministry of the Treasury, the presidency of the Central Bank, and the directorship of the CVF have all been under the aegis of representatives of the major finance groups.[92] During the Caldera administration (1969–74) the minister of communications was Oscar Machado Zuloaga, the don of the Zuloaga finance group, which owns the Caracas Electricity Corporation. As was particularly evident during the same regime, past Fedecámaras presidents have provided business with access to the state's oil revenue: Carlos Guillermo Rangel was director of the Venezuelan Development Fund and Pedro R. Tinoco, Jr. served as finance minister.[93]

Fund raisers for COPEI during the 1968 campaign were also strategically placed: Alfred Rodríguez Amengual, a director of the Popular Homes Corporation (VIPOSA), was minister of housing; his brother, Gustavo, headed the Simón Bolívar Center; and Haydée Castillo, previously the director of the Mendoza group's economic research staff, became the minister of development.[94]

The development of a bureaucratic bourgeoisie stratum by the 1970s indicated that the Venezuelan ruling class was creating another tie to the state apparatus: government officials, by providing favors to business groups while in office, would be rewarded by receiving capital ownership in corporations after they left public employment. Llovera Páez and Rómulo Fernandez, two retired military officers who were in the civilian government in the 1960s, became major capitalists as a result of their offices.[95] Luis Melo Quintero, legal counsel for the CVF and later attached to the BIV, today owns at least 40 percent of the shares in the Arauca Insurance Company. The national Congress (dominated by AD even during the Caldera administration) discovered that considerable corruption in the Ministries of Public Works, Communication, and Agriculture and Livestock, as well as in the IVP, had existed during Leoni's tenure in office.[96] Even in cases where no direct business-government ties exist, the mixed company status of such agencies as the CVF and the BCV allows the national capitalists to exercise their ownership prerogatives to influence economic policy.

State Capitalist Developmentalism

The expansion of state activity in the promotion of stable economic growth is one of the most salient features of the Venezuelan social formation. The creation of autonomous agencies, state holding companies, state investment funds, and so on has had the double function of opening up the agricultural sector to commercial exploitation and extending the industrialization process and complementing and extending opportunities for the private sector, especially national capital. Throughout recent Venezuelan history, the state has been a necessary, but not independent, force in the drive for development. Its role has been principally defined by the interests of the emerging national bourgeoisie, in which the state assumes all the risks in initial investments, especially in low-profit areas, and the capitalists take advantage of the resulting opportunities. The preponderant weight of the state activity in the economy should not obscure the private ends that are served.

While commercial agriculture and livestock production has been increased in the last 25 years, land concentration has remained relatively unchanged.* As is argued, Venezuela is not now, nor ever has been, a dual society. The strong ties between urban financial capital and the countryside preclude the possibility that sectoral reforms of agriculture can withstand the process of capital accumulation. By 1967, 50 percent of the aggregate family income of the land reform settlers was from wage employment[97] and, by 1970, 30 percent had abandoned their family plots.[98] Due largely to the monopoly of private and government credit by the large commercial farmers, the rural settlers have been proletarianized by their more affluent neighbors.

The dependent nature of the Venezuelan bourgeoisie required that the state had to assume the initiative in formulating the industrialization strategy. During the 1960s, in an attempt to extend the import-substitution process begun in the previous decade, Pérez Jiménez' free trade pact with the United States was revoked and import quotas and exchange controls were instituted. As the growth opportunities provided by the trade barriers were exhausted, the Rafael Caldera and Andrés Pérez regimes turned to a new strategy. State and mixed enterprises continued to be promoted in basic industry, while in the private sector the diversification of manufactured exports was financed through the autonomous institutes.[99] The two major economic policies of the 1970s, Venezuela's entry into the Andean Pact and the nationalization of the petroleum industry, grew out of this strategy. Neither could have occurred without the development of a home market, a nationalist bourgeoisie, and a

---

*The index of arable land concentration as measured by the Ministry of Agriculture census shows practically no change between 1950 (0.942) and 1971 (0.909).

capitalist state; these three preconditions created a finite set of social needs and the possibilities for meeting them. Nonetheless, the two events were the culmination of processes set in motion by prior government policies.

In summary, during the first AD government, in addition to the establishment of OPEC in 1960, oil taxes were again increased and the CVP was reinstituted. In 1966, the U.S. oil companies began to lose political support among Venezuelan businessmen after the U.S. government reimposed import quotas on Venezuelan oil. The Leoni administration empowered the CVP to distribute oil within the home market, and, in 1967, charged it with the responsibility for administering the new, more effective, service contracts system of fiscal control of the oil industry. This arrangement provided the Venezuelan state for the first time with inside information on the real profit rates of the oil corporations. Equally important ideologically, under the new system the state enjoyed subsoil rights over the oil fields, which were merely leased to the foreign capitalists. The hand of the nationalists was further strengthened in 1967 when the oil corporations, in contesting the law, were defeated in the Venezuelan courts. The same year saw the first effective functioning of OPEC.

Riding the wave of nationalist sentiment, Rafael Caldera, in his 1968 campaign, took the position that the existing oil concessions should revert to the state after 1983; by 1971, the foreign-owned natural gas industry was nationalized and the oil reversion law had been passed. The measure called for the immediate government takeover of all the foreign concessions not currently being exploited.[100] That year the reference prices on crude oil, which began to be fixed by the government for tax purposes, increased the state's share of oil export profits to 80 percent.[101] Finally, as it became apparent that the oil corporations had adjusted their production levels so that the known oil reserves would run out at the same time as their concessions, Caldera signed a decree that empowered the oil minister to stop petroleum production and that required the foreign companies to submit their projected production levels to the government for review and approval.

Since the government could not really force the foreign capitalists to invest, however, nationalization began to be discussed as the only remaining alternative. This became a real possibility in 1973. Due to the role of the U.S. government in supplying arms to Israel, the governments of the Arab member nations of the OPEC cartel imposed an embargo on their oil exports to the United States. By 1975, the price of Venezuelan crude had jumped from $2 to $14 per barrel and the annual public budget had increased from Bs.14 billion to Bs.42 billion. This enormous reserve made the Venezuelan government relatively immune to any financial leverage that might be applied by the oil corporations or their governments.

By the time of the 1973 presidential elections, the nationalization momentum was difficult to resist. The eventual victor in the presidential campaign,

AD candidate Carlos Andrés Pérez, stated unequivocally in the party's platform, *Acción de Gobierno*:

> Since there remain few years until reversion to the Nation of the greater part of the present concessions, the private companies are maintaining at minimum levels their exploratory activities and we run the risk that our industry, due to failure to incorporate new techniques and the absence of appropriate investment and maintenance, will rapidly deteriorate and we shall find ourselves at the time of delivery of the concessions with outworn equipment and obsolete technology. For these reasons, it appears impossible to wait until 1983 before the State assumes the full management of the petroleum industry. In order to maintain the present industry in satisfactory condition and to carry out other aspects of the petroleum program presented, it would be prudent, as a possible alternative, that we proceed in the immediate future to nationalize, which ensures our sovereignty in the industry and that it will arrive at new formulae for the participation of foreign companies in those aspects in which we need their technical resources, their financing or their commercialization ability. These formulae are being put into practice in other important petroleum countries and other new ones can be developed to satisfy our aspirations and our interests.[102]

In January 1976, the Pérez government officially nationalized, with compensation, the foreign-owned iron and petroleum company assets in Venezuela.

In analyzing the social forces responsible for the advocacy of state capitalist developmental policies over the last 15 years, one fact in particular stands out: the lack of independence and initiative displayed by Venezuelan finance capitalists. Fedecámaras, the representative organ of the nation's most powerful economic groups, has initiated none and opposed most of such policies. In the 1960s, since it was interested primarily in valorizing existing capital, Fedecámaras criticized the government for its creation of artificial import-substitution industries producing high-cost, low-quality goods. It charged that the program constituted, in effect, a disguised public employment program. In the late 1960s and early 1970s, however, when the need to diversify Venezuelan exports (given the constraints on the home market from the existing income distribution) was apparent, Fedecámaras put up the most intransigent opposition of any such Latin American business pressure group toward proposals for a regional free trade area (LAFTA) or a customs union-subregional economic integration (Andean Pact).* Because of their own dependence on state protec-

---

*Interests within Fedecámaras delayed Venezuelan entry into LAFTA from 1967 until 1969 and successfully fought the government's participation in the Andean Pact from 1968 to 1973.

tion and subsidies, the Venezuelan monopoly bourgeoisie were high-cost, inefficient producers who feared the competition of cheap imports.*

Fedecámaras did the least of any organization to press for the nationalization of oil. During the two AD administrations, it virtually echoed the criticisms of the government's oil policies made by the U.S. oil corporations. As nationalization became an increasingly considered policy option during the Caldera regime, Fedecámaras changed its tactics on the issue from outright principled opposition to objecting to the speed at which the nationalization process was proceeding, raising the issues of the lack of oil marketing outlets and the scarcity of trained Venezuelan management. In a backhanded way, however, the business organization did contribute to the eventual state takeover of the industry; by resisting so successfully any and all attempts by the Congress to impose greater rates of taxation on corporate profits, it eliminated the most viable alternative source of government revenue.

Those social forces that made positive attempts to bring about the nationalization were from several classes. Clearly, the nationalist elements of the state petty bourgeoisie played a crucial role. The major oil propagandist and author of much of the petroleum industry legislation, Pérez Alfonzo, was particularly important as were some of the nationalist economists in the planning bureaucracy who had been influenced in the 1960s by the staff of the United Nations Economic Commission on Latin America. Factions of the small and medium-sized import substitution bourgeoisie that were less tied to foreign capital and markets adopted positions antagonistic to the foreign oil corporations. They formed two business pressure groups, the Venezuelan Executives Association and Pro-Venezuela, which tended to oppose the proimperialist political positions taken by Fedecámaras. Much of the nationalist petroleum policies originated from the legislation of the AD-controlled Congress. Within the legislature the URD and the MEP representatives unsuccessfully cosponsored an oil nationalization bill in 1971. The union bureaucracy provided very little nationalist leadership.[103] However, the wave of rank and file strikes that occurred from March 1969 to October 1970, which AD and COPEI tried to suppress, established a social context that may have been conducive to the passage of the 1971 oil legislation. Also, the oil workers effectively ensured the enforcement of the measures. Following the passage of the oil reversion bill, the petroleum companies responded by drastically cutting production levels, which entailed laying off workers; the Federation of Petroleum Workers (Fedepetrol) threatened to go on strike unless the previous production levels were maintained.

---

*The CTV also opposed the entry due to fear of job loss caused by low wage imports.

In spite of their role in the oil nationalization, the major beneficiaries of this process, as indicated by the CVF allocation of investment credit between 1958 and 1967, have been those finance groups that have received the bulk of the state-appropriated oil revenue. The two sugar and cement firms of the Vollmer-Zuloaga group were the most favored. The large paper and ceramics corporations controlled by the Mendoza group were the second most heavily subsidized. The three enterprises (primarily in paint) of the Montana faction ranked third, followed by Boulton, Phelps, Sosa Rodríguez, Blohm, and Brillenbourg.[104]

Similarly, the FIV has aided the recent ascendence of the Western group, which was formed around the axis of the Western Mortgage Bank and the Andean Insurance Company.[105] The Caribbean Cement Corporation, owned by the principal members of the group (Pedro Tinoco, Jr., Enrique Delfino, and Ciro Febres Cordero), has been particularly fortunate. Following Febres Cordero's generous contributions to the 1973 AD campaign, Caribbean Cement has received Bs.300 million from the Investment Fund, more than 12 times the amount of the corporation's paid-in capital of Bs.20 million.[106]

However, the Caribbean Petrochemical Corporation, owned principally by Tinoco, Delfino, and Domingo Mariani (all of the Western group), presents what is perhaps the exemplar of the future direction of Venezuelan state capitalism following the oil nationalization. The corporation's new oil refinery in Costa Rica, which will have an initial capacity of 100,000 barrels per day, will be used to penetrate the Central American market. The plans for the plant's installation were initially set up by Tinoco when he visited that country in his official capacity as president of the Public Administration Integral Reform Commission. The project brings together all the elements of the current period of Venezuelan capitalist development; the capital will be provided cheaply by the Venezuelan Investment Fund and the crude oil will be available at state-subsidized prices from PETROVEN (Petroleum Company of Venezuela), a new state petroleum enterprise established under a provision of the 1976 nationalization.[107]

CONCLUSION

A number of general propositions emerge from this sketch of the history of Venezuelan capitalist development that are perhaps useful points of departure for research in other Third-World countries.

The existence of a *comprador* (import-export) bourgeoisie is not incompatible with the development of national capitalism. On the contrary, under conditions of strong state promotion there exists the possibility of conversion from one bourgeois type to another without any basic upset of the social structure. The conversion largely involves the transfer of capital accumulated in one sector to another, tying the economy as a whole into a web of interlocked capitals.

The bourgeoisie in the Third World, though it is dependent on foreign capital, and in no consequential and sustained fashion promotes national expropriation of major imperial firms, is capable of taking advantage of such changes promoted by other classes and turning them to its own account.

As a corollary, petty bourgeois and intermediary bureaucratic strata may initiate fundamental changes in the pattern of ownership—displacing foreign capital—but are incapable of producing a society in their own image: either large-scale foreign enterprises cannot be broken up into small firms or the statified firms serve the ends of large-scale private interests.

Throughout the process of capital accumulation, the main instrumentality is the state. Control and direction of the state become the central preoccupation of all aspiring bourgeois forces. Control over the state will shape the degree to which one faction or another of the bourgeois class will advance. The direction of economic advance is largely dependent on which faction is in control of the state. The struggle over government policy is a reflection of the extreme dependence of the private sector on the state.

The policies of particular governments frequently clash with the immediate interests of the local bourgeoisie, especially in transition periods when the leading sector of the economy is changing, for example, from agriculture to industry. The linkages between economic sectors and between national and foreign capital inhibit class formations from innovating. Within the same mode of production, the changes in sectoral predominance will be initiated by the state and yet may be initially opposed by the ultimate beneficiaries. Capitalist modernization will be initiated outside the capitalist class by its political leadership and yet will ultimately broaden the scope of its activity, as was the case with the industrialization program, the agrarian reform, and the nationalization of petroleum. Hence, there is no identity of interest in the short run between the capitalist class and/or parts of the political leadership; only in historical perspective does the convergence of policy and outlook of the two occur.

Differences between authoritarian and democratic political regimes on welfare and redistributive measures are not significant. The historic experience reveals that only marginal changes in income, class, and status will occur. The major shifts occur within the ruling class. Insofar as centralized military regimes block the ascendence of emerging capitalist groups, they lead to greater instability than democratic regimes. Insofar as democratic regimes do not hinder the process of capitalist accumulation, political opposition is tolerated, indeed is involved in the political decision-making structure.

The long-term trend is for nationalist and populist regimes to collapse or be overthrown. The alternative is an accomodation with the social classes central to the process of capital accumulation. Insofar as those classes have linkages with outside foreign capital, the regime will restrict its nationalist measures to selective nationalization; insofar as the classes are linked across economic sectors, it will vitiate attempts at internal sectoral change.

Changes in the role of the state reflect the internal changes in the nature of the ruling class and its capacity to handle its "fit" in the world economy. The decline of agriculture and the rise of industry lead directly to a more activist state. The growth of new export commodities and their subsequent regulation or control will extend the activities of the state. The growth of the state will in turn propel new strata into political power that will share with the bourgeoisie the direction of the economy but conflict over the specific allocations. The cleavages and consensus between bureaucratic and entrepreneurial capital dominate the political arena—defining the boundaries within which reformist and conservative politicians vie.

Over and above the nationalist commitments, the imperatives of capitalist accumulation dictate the personnel of the new state emerging with nationalization. The flow of investments and funding is from the state to the private sector, limiting the scope and depth of planning. Technocratic functionaries appear most influential in times of crises or early periods of new exploitation when private capital is inadequate for the tasks at hand. In the long run, both the technocrats and their state apparatus become enmeshed in the task of becoming their own gravediggers—the private sector gradually replaces the more lucrative ventures, leaving the state the more costly and least profitable enterprises.

NOTES

1. Domingo Alberto Rangel, *Capital y Desarrollo,* vol. 1 (Caracas: Imprenta Universitaria de Caracas, 1969), p. 272. In Maracaibo, the major trade depot in the nineteenth century, 60 to 80 percent of the coffee exports were controlled by the foreign houses, while in Caracas, second in importance, they controlled 70 percent of the cacao trade.

2. Ibid., p. 274.

3. Ibid., p. 308. By 1920, industrial capital only represented between 2 and 3 percent of the national real fixed capital.

4. Ibid., p. 190. In 1910, 30 percent of the annual real capital investment was in agriculture; by 1920, it was 18 percent.

5. Ibid., p. 255.

6. Ibid., p. 257. Located at the seat of the federal capital and selected as the bank that would hold state deposits from the collection of taxes (customs duties), by 1912 the Bank of Venezuela was issuing 60 percent of all the country's bank notes and held 60 percent of its gold reserves and bank deposits.

7. Ibid., p. 210. In 1900, urban construction comprised 25 percent of all national real fixed investment, was the biggest sector by 1910, and constituted 85 percent by 1920.

8. Domingo Alberto Rangel, *Capital y Desarrollo,* vol. 2 (Caracas: Imprenta Universitaria de Caracas, 1970), p. 117.

9. Ibid., pp. 152–53.

10. Ibid., p. 133. Venezuela's national income doubled between 1922 and 1929.

11. Ibid., pp. 166–68. From 1920 to 1940, construction received twice as much capital as either agricultural machinery or manufacturing and more than five times the amount applied to artisan production.

12. Ibid., p. 169.

13. Ibid., p. 132.

14. Ibid., p. 228.

15. Ibid., p. 269.

16. Ibid., p. 245.

17. Ibid., p. 269.

18. Franklin Tugwell, *The Politics of Oil in Venezuela* (Stanford, Calif.: Stanford University Press, 1975), p. 20. In the years 1920 to 1943, the absolute amount of such royalties, taxation, concession sale proceeds, and so on doubled each ten years; between 1943 and 1947, government oil revenue quadrupled.

19. Rangel, op. cit., vol. 2, pp. 317–20. Between 1936 and 1953, in a period when the bolivar's exchange rate with the dollar increased from Bs.5.20 to Bs.3.35, industrial production expanded in value from Bs.119 million to over Bs.3 billion. Within the same time span, the average number of workers per firm and the profit per worker doubled, while the value of fixed industrial capital increased from Bs.119 million to Bs.964 million.

20. *Pre-Diagnóstico de la Economía Nacional,* Segundo Borrador (Caracas: Centro de Estudios de Desarrollo Económico y Social[CENDES], 1975). Overall, by 1950, manufacturing would represent only 10 percent of the gross national product.

21. Rangel, op. cit., vol. 2, p. 147; David E. Blank, *Politics in Venezuela* (Boston: Little, Brown, 1973), p. 37; and *Economic and Social Progress in Latin America: 1974 Annual Report* (Washington, D.C.: Inter-American Development Bank, 1974), p. 423.

22. Rangel, op. cit., vol. 2, p. 113. An idea of the less than totally indigenous character of the postwar industrialization can be gained by noting the changing level of U.S. investments in Venezuela: $245.3 million in 1929, $262 million in 1940, and $993 million by 1950.

23. S. Menshikov, *Millionaires and Managers* (Moscow: Progress Publishers, 1973).

24. Rangel, op. cit., vol. 2, p. 317.

25. For a more detailed analysis of the military in this period, see Winfield J. Burggraaff, *The Venezuelan Armed Forces in Politics, 1935–1959* (Columbia: University of Missouri Press, 1972).

26. See E. Baloyra, "Oil Policies and Budgets in Venezuela," *Latin American Research Review* 9, no. 2 (Summer 1974), pp. 28–72.

27. Tugwell, op. cit., p. 42.

28. Ibid., p. 43.

29. Rangel, op. cit., vol. 2, pp. 319–20.

30. Tugwell, op. cit., p. 45.

31. Jorge Ahumada, "Hypothesis for Diagnosing Social Change: The Venezuelan Case," in *A Strategy for Research on Social Policy,* ed. Frank Bonilla and José Silva Michelena (Cambridge, Mass: MIT Press, 1967), p. 11.

32. Memoría, Banco Central de Venezuela, 1959.

33. International Bank for Reconstruction and Development, *The Economic Development of Venezuela* (Baltimore: Johns Hopkins Press, 1961), p. 90.

34. See Thomas E. Weil, et al., *Area Handbook for Venezuela* (Washington, D.C.: American University, 1971). This represents an increase of 4.6 percent since 1950.

35. See International Bank for Reconstruction and Development, op. cit., p. 95. This constituted a considerable diversification. The percentage of manufacturing values in food, textiles, and beverages decreased from 50 percent in 1950 to 33 percent in 1958, and the share of manufactures in consumer goods dropped from 71 to 59 percent during the Pérez Jiménez regime.

36. Ahumada, op. cit., p. 10. Both grew at 20 percent.
37. International Bank for Reconstruction and Development, op. cit., p. 85.
38. Ibid., p. 86.
39. Ibid.
40. Domingo Alberto Rangel, *Capital y Desarrollo,* vol. 3 (Caracas: Editorial Fuentes, 1972), p. 57.
41. Ibid., p. 59.
42. Ibid., p. 60
43. International Bank for Reconstruction and Development, op. cit., p. 107.
44. Ibid., p. 90. Tariff levels on intermediate and capital goods averaged only 5 percent and those on consumer durables 15 percent. The average tariff on goods actually imported was a moderate 20 percent.
45. Ibid., p. 106. Foreign exchange reserves decreased $360 million in both 1958 and 1959 and would have declined much farther had the provisional government not borrowed considerable sums from the U.S. government.
46. Ibid., p. 109.
47. Direccíon de Cultura, Instituto de Investigaciones de la Facultade de Ciéncias Económicas y Sociales, *La Evaluación de la Inversión del Ingreso Fiscal Petrolero en Venezuela* (Caracas: Universidad Central de Venezuela, 1968), p. 238.
48. Philip B. Taylor, Jr., *The Venezuelan Golpe de Estado of 1958: The Fall of Marcos Pérez Jiménez* (Washington, D.C.: Institute for the Comparative Study of Political Systems, 1968).
49. See Robert Alexander, *The Venezuelan Democratic Revolution* (New Brunswick, N.J.: Rutgers University Press, 1964).
50. See volumes for the appropriate years of the U.S. Department of Commerce, "U.S. Foreign Direct Investments (m$) for Selected Countries and Industries," *Survey of Current Business.* Figures on U.S. direct investment in Venezuela reflect this trend. Although U.S. capital continued its penetration of Venezuelan manufacturing, commerce, and other sectors between 1957 and 1959, its investments in utilities dropped to pre-1956 levels.
51. Américo Martín, *Los Peces Gordos* (Valencia: Vadell Hermanos, 1975), p. 208. This allowed the Caracas Electricity Corporation, owned by the Vollmer-Zuloaga group, to increase its rate of profit on total invested capital in the 1960s to nearly 27 percent.
52. Alexander, op. cit., p. 167. For example, state financing of large ranchers and farmers increased substantially during the provisional regime; in 1957, the Agricultural and Livestock Bank provided Bs.50 million in credits; within two years the figure had become Bs.200 million.
53. See John D. Martz, *Acción Democrática: Evolution of a Modern Political Party in Venezuela* (Princeton, N.J.: Princeton University Press, 1966).
54. Héctor Malavé Mata, *Dialéctica de Inflación* (Caracas: Universidad Central de Venezuela, 1972), p. 312.
55. Ibid., p. 81. Of course much of the variance in gross national product growth rates is due to changes in the value of its second major component, oil itself.
56. Héctor Silva Michelena, "Proceso y Crisis de la Economia Nacional, 1960–1973," *Nueva Ciéncia* 1, no. 1 (1975): 128.
57. CENDES, op. cit.
58. Ibid. By 1973, 80 percent of Venezuela's imports were in primary and intermediary capital goods for agriculture, mining and manufacturing.
59. Miguel A. Falcon Urbano, *Desarrollo y Industrialización de Venezuela* (Caracas: Universidad Central de Venezuela, 1969). Fifty percent in 1961 and 40 percent in 1966.
60. Weil, et al., op. cit. In 1973, agriculture's share of gross national product was approximately 6 percent, while 22 percent of the work force was employed in this sector. Mining and petroleum, although together they produced 23 percent of the gross national product, only employed about 2 percent of the work force.

61. CENDES, op. cit. In 1973, an estimated 15 percent of the national income was drained off by foreign capital: 10 percent for production factors and 5 percent for royalties, patents, services, and so on.

62. Industry is able to absorb only 10 to 15 percent of the yearly additions to the work force: nominal, or official, unemployment ("currently looking for work") ranged from 13 to 5 percent between 1961 and 1973 (Inter-American Development Bank, op. cit), while real unemployment ("percentage of the economically active work force") has been estimated to have been about 33 percent in 1970 (CENDES, op. cit).

63. Rangel, op. cit., vol. 3, p. 351. Total consumer credit comprised 12 percent of the Venezuelan gross national product in 1967.

64. Ibid., p. 350. As early as 1959, 72 percent of all such sales in Caracas were made on credit.

65. Ibid. In 1962, such loans totaled Bs.186 million; by 1967, the amount was approximately Bs.900 million.

66. Weil, et al., op. cit.

67. Michelena, op. cit., p. 128.

68. Weil, et al., op. cit.

69. Blank, op. cit.

70. CENDES, op. cit. In 1971, the average after-tax net profit rate on fixed capital for large manufacturing firms was 38 percent. This compares very favorably with the net after-tax profit of 32 percent on total invested capital in the petroleum industry for the same year.

71. Blank, op. cit.

72. CENDES, op. cit.

73. Málavé Mata, op. cit., p. 322.

74. Juan Pablo Pérez Alfonzo, *Petróleo y Dependencia* (Caracas: Sintesis dos Mil, 1971), pp. 24–25.

75. Michel Chossudovsky, *Pobreza y Marginalidad en Venezuela* (Primera Versión) (Ottawa: University of Ottawa, 1975), p. 331.

76. CENDES, op. cit.

77. Ibid.

78. Ibid.

79. Ibid.

80. Chossudovsky, op. cit., p. 198. The other six countries are Argentina, Brazil, Colombia, Costa Rica, Mexico, and Panama.

81. Ibid., pp. 194, 201, 337.

82. Michelena, op. cit., p. 22. Theoretically, this is surplus value of product per worker divided by necessary value of labor in product; empirically, this rate is estimated by the value added of a worker's daily product divided by the worker's daily wage.

83. CENDES, op. cit. Monopolies were 7.1 percent of the total number of industrial firms by 1971.

84. Chossudovsky, op. cit., pp. 38–39. Assuming an average family size and number of wage earners per family, the income required to provide the minimum necessary food, health, housing, and clothing.

85. Ibid., p. 220. In 1975.

86. Ibid., p. 298.

87. Ibid., p. 300.

88. Ibid., p. 301.

89. Robert Clark, Jr., "Economic Integration and the Political Process: Linkage Politics in Venezuela," in *Venezuela 1969: Analysis of Progress,* ed. Philip B. Taylor (Houston: University of Houston, Office of International Affairs, 1971).

90. Inter-American Development Bank, op. cit., p. 434.

91. Weil, et al., op. cit.

92. Rangel, op. cit., vol. 3, p. 386.
93. Allan-Randolph and Brewer-Carías, *Cambio Político y Reforma del Estado en Venezuela* (Madrid: Editorial Tecnos, 1975), p. 265. Tinoco has been very closely associated with U.S. capital: previous to his positions in government he had been president of the Commercial and Agricultural Bank (Chase Manhattan) and in 1965 was legal counsel for Bethlehem Steel Corporation in Venezuela. He was the architect of the 1960's Fedecámaras' doctrine and later formed the developmentalist group, which he allied with the Pérez Jiménistas and the Capriles publishing monopoly.
94. Ibid., p. 103.
95. Ibid., p. 121.
96. Martín, op. cit., pp. 123–25.
97. Weil, et al., op. cit.
98. Blank, op. cit.
99. CENDES, Table 6, op. cit. By 1971, two years before the big increase in oil prices, over 45 percent of the government's capital expenditures were in financial investments.
100. "Venezuela: Caldera's Delicate Balance," *Latin America,* August 6, 1971. This amounted to five sixths of the total existing concessions.
101. H. J. Maidenberg, "Setback in Venezuela," New York *Times,* January 17, 1971, sec. 3, p. 1.
102. Carlos Andrés Pérez, *Government Action* (Caracas: Editorial Arts, 1973), p. 23.
103. Blank, op. cit., p. 234. In the mid-1960s, 56 percent of the nation's labor officials were opposed to the nationalization of oil.
104. Rangel, op. cit., vol. 3, p. 388.
105. Martín, op. cit., pp. 164–65.
106. Ibid., p. 170.
107. Ibid., pp. 166–67.

CHAPTER 2

NATIONALIZATION AND CAPITALIST DEVELOPMENT

INTRODUCTION

The emergence of durable regimes that proceed to nationalize profitable foreign-owned enterprises is one of the central developments in the Third World today. The implications of these developments are significant, as they have an effect on class formation and particular patterns of economic activity. Having noted the general growth of statism, there is a need to specify the types of social formations that emerge from the initial acts of state definition.

The approach used here begins by distinguishing several types of nationalization and the roles that they play within the larger development problematic according to the regime that dominates the social formation. Within this larger discussion, nationalization is considered within a particular type of regime: state capitalism, the type that most closely approximates the Venezuelan approach. The discussion focuses on the factors contributing to nationalization, and specific thrust of the strategy that accompanies it, and the implications in terms of capitalist class-state relations.

In addition, the Venezuelan nationalization process must be considered within the larger international setting in order to account for the timing of the event. For while internal conditions have for some time been favorable, it was a combination of internal class developments and international political developments that finally led to the taking over of the oil resources. The final section considers the institutional and social class impediments toward the maximization of growth and social equality through the nationalization. Here analysis goes from the international level back to the national, tracing the behavior of the ruling class, state bureaucracy, and the political process as they affect the allocation of the surplus and the execution of development projects.

ASPECTS OF NATIONALIZATION IN THE THIRD WORLD

The plurality of models of development and the varying sequences should not obscure the general policy lines that different state capitalist regimes follow. For some regimes, state ownership of foreign firms becomes largely an instrument to finance the advancement of private sector activity, channeling funds through private entrepreneurs within the country and to foreign capital on the outside. With other regimes, the tendency is to use the nationalized sector as a springboard to advance state enterprises in a variety of industrial, financial, and commercial activities, including major sectors of the economy, either displacing or competing with the private sector. These two polar subtypes do not exhaust the variants, which can include an infinite variety of mixes, but rather point in two possible directions in which state capitalism can move (see Table 1).

The Postnationalization Problematic: Surplus Disposition

With the nationalization of profitable key basic industries, the surplus that previously was siphoned off by foreign entrepreneurs is available for investment within the national economy. The flow of funds from the state thus becomes the central issue in the postnationalization period: the control of the commanding heights of the economy is but the first moment in determining the social nature of the regime. Numerous variants of the social formation can emerge, depending on the role that the state will play in organizing and directing the derivative, related and even unrelated, industries that can be generated by the surplus. In turn, the capacity of the state to stake out an autonomous path of expansion—to reproduce an accelerated expansion of state enterprises—is dependent on a number of factors: the preexisting social structure in which the nationalization takes place; the permeability of the state party; the manner in which the state party comes to power and the manner by which expropriation takes place; the nature of the social polarization that precedes the nationalization; and the degree to which national capital is integrated directly or indirectly with foreign capital affected by the nationalization.

Class Structure and Nationalization

It is rarely the case where private national capitalist groups take the initiative and organize the effort to nationalize profitable foreign private capital. Foreign markets, imports, inputs, credits, licensing agreements, management contracts, royalty payments, and investment ties create a common bond between firms and operations in the periphery and the metropolis. These

TABLE 1
Typology of State Capitalist Nationalization
(surplus disposition by sectors)

Role of Nationalized Firms	Agriculture	Industry	Societal Consequences
Predominantly as channels for private national capital	Agribusiness Entrepreneurial farmers	Private national industrial and financial capital State exploitation of raw materials Mixed firms in derivatives	Concentration of private capital Expansion of bureaucratic strata divorced from production Proletarianization of agrolabor Increased inequalities
Predominantly as base for formation of mixed private-state enterprises	Entrepreneurial farmers Agribusiness Cooperatives	Mixed firms-state ownership in basic industry Licensing and management contracts Private sector	Concentration of income within industrial state-private capital sectors Expansion of role of bureaucracy in economy Proletarianize labor Increased inequalities
Predominantly as base for state enterprises	State farms Cooperatives	State firms in basic industry State control of commerce and banking Mixed firms in manufacture Service contracts	Displacement of foreign-private elite by state capitalists Income concentrated among state elites Proletarianized agrolabor force Large surplus urban labor force

linkages inhibit sustained and consequential nationalist capitalist action against the foreign-owned enclaves. Before nationalization, growth of national industries is largely financed by the surplus that accrues to the state from tax payments and/or partial ownership. The coexistence of national industries and foreign-owned enclaves persists for a considerable period without any severe strain or conflict. The industrial sector surrounding the enclave is embedded largely in a network of state aids that facilitate the importation of semielaborated goods for assembly and finishing. The semicommercial, semiindustrial nature of the industrial importers-manufacturers makes this class less than a protagonist of nationalist politics.

The growth of peripheral nationalism, then, is a noncapitalist political form within a nationalist capitalist society. The possibility of this noncapitalist political form taking power and appropriating the surplus for the national economy causes a shift within the national capitalist class. Because its ties are mainly with metropolitan industrial firms and not directly with the enclave, the national capitalist class can adapt to the nationalist impulse to expropriate the dynamic enclave sector. The problematic from the view of the national capitalist class is how to seize the surplus accruing to the national state without endangering external linkages. To the degree to which a differentiated class structure exists in which a national capitalist class of fairly defined dimensions exists, there is a high probability that the nationalization of foreign enclaves will become the instrumentation of accelerated national private capitalist expansion. Nationalization of foreign-owned dynamic sectors inserted into the advancing capitalist formation serves to concentrate further national private capital and expand capitalist social relations.

Conversely, where there is a very weakly differentiated capitalist society, with a private sector that shows little capacity to absorb and convert the surplus into capital investments, and, more important, where the sociopolitical weight of the nationalist private capitalist class in the state apparatus is relatively light, there is a high probability that the state will become the engine of industrial expansion. The possibilities of long-term state expansion are inversely related to the growth of a powerful private sector. Insofar as the state enterprises become linked to the private sector, there will be a tendency to lower the cost of the inputs from the state sector and increase the costs to the state firms. The private capitalist looks to the state as the surplus collector and expeditor, not as a competitor or director. The extreme privatized tendency within the postnationalist society is to reduce the state to the role of providing investment funds for private investors (indeed, the very financial channels through which the funds are funneled are to be in the hands of private banks). The long-term effects are to expand capitalist property relations and proletarianize sectors of the labor force while displacing precapitalist social strata (small farmers, artisans, and so on) toward the urban subproletariat. Surplus, at the service of the national private captialist class, increases the market for industrial and consumer exports from the metropolis, generating pressures from inside the metropolis (among industrial exporters) against the enclaves' political reach in the foreign policy bureaucracy. Simultaneously, the strength of the national private sector in the periphery serves to limit the external reach of the nationalized sector, tying it to continuing relations with the foreign sector (via commercialization and so on).

The importance of liaison groups within the national class structure in limiting and defining not only the internal dynamic of state entrepreneurial activity but their external relations as well cannot be underestimated. The focus of the left on the nationalization issue, per se, rather than on the question of nationalization for whom—that is, the focus on the external dependency to the exclusion of internal class relations—leads to the strengthening of the national capitalist class, increasing its capacity to operate, and perhaps stabilizing its expansion within a longish period of time. As noted above, the issue of nationalization is not one that was intially pushed by the capitalist class, but, when the issue is joined and it is ensured of virtual success, the central issue for the capitalist class becomes one of securing unlimited access to the surplus and its conversion into private investment capital. Upon possession of the surplus, the private national sector is free to establish a whole new series of ties with foreign capital wholly integrated in the economy and not set apart as an alien enclave. The incorporation of foreign capital in the industrial, commercial, and service sectors coincides with the broad expansion of national capital. This transformation marks the definitive end of national popular politics and the clear point of departure for class politics.

The Permeability of the State Party

The preexistence of a highly differentiated class structure with a developed capitalist class is a necessary but not sufficient condition for the diversion of surplus from state to national private capital expansion. The central point here is access to the state; and the preconditions for entree are given in the political formation that organizes the taking of state power. The existence of a faction of the national private capitalist class within the political formation that controls the government is a prerequisite for influencing the direction of executive policy. The direct presence of a private capitalist faction within the executive, then, lends weight and magnifies the outside influence exercised by the class as a whole or its representative bodies. The internal contest, then, between the statist and private entrepreneurial orientation of the government is an uneven match. The globality of private entrepreneurial involvement (mass media, economic levers, and pressure groups) encircles the state sector and limits the scope of its operation, dictates the economic priorities on agenda, and ultimately emasculates efforts at autonomous state economic activity. Without effective political representation, the private capitalist class can only make its presence felt from the outside.

The differences between the Peruvian and Venezuelan examples of state capitalism are not found in the existence or nonexistence of a preexisting capitalist class but in the degree of permeation of the hegemonic political force. In Venezuela, there are strong interlocks between the capitalists and the dominant political party; in Peru, the virtual absence of the capitalist class from the dominant military regime favors the expansion of the state sector. The limited access and precarious linkages between capitalists and state permit the statist sectors to multiply and expand the state sector, reproducing the original state conquest in a multiplicity of economic areas. The nondemocratic political forms provide few obligatory ties between political representative and entrepreneur: the financial nexus that emerges out of expensive electoral charades so present in Venezuela is absent in Peru. The existence of a relatively nonpermeable state apparatus favors the statist sectors of the state capitalist regime and provokes limited confrontations and competition between state capitalists and the national private bourgeoisie. The political formation mediates class power rather indirectly, reproducing capitalist class relations (within and between state enterprise), but without representing private capitalist interests.

The Path to Power and Expropriation and Their Effects on Postnationalization Surplus Acquisition

The process of private surplus acquisiton depends on the existence of a well-defined national capitalist class that permeates the party that is politically dominant within the state structure. The process of penetrating ascending political formations is facilitated by prolonged electoral exercises that depend on extensive media exposure and high financial costs. The electoral contest, which, through massive campaigns, puts a premium on financial resources, creates a set of political obligations between entrepreneur and aspiring politicians that facilitate long-term access and convergence of interest. The capacity to draw on financial support from the business community itself reflects the socioeconomic disposition and personal aspirations of the politician: the conversion of political into business success. The mutual manipulation between politician and business is not exempt from frictions, though one is likely to find, over time, a tendency for roles to fuse. The whole ensemble of political roles, economic obligations, and social aspirations linked to the electoral process facilitates the gravitation of policy decision making toward the national capitalist class, which becomes a central figure in directing the nationalization process.

Polyclass political movements that draw their electoral base from among the poor and their directorship from the business community provide the central instruments exchanging foreign capital control for national. At the same time, contradictory forces within the polyclass movement continue to exercise pressure from below but increasingly from the outside toward greater social consumption. In the middle are the state bureaucratic sectors that define the political economic project as neither private expansion nor consumption but state development. Within an open electoral arena, without the protection of an insulated state apparatus, the state bureaucratic sectors fight a rear guard action, throwing up enterprises, plans, and projects and then gradually ceding ground before the encroachments of the private sector. The electoral polyclass parties with their enormous and heterogeneous clientele of potential and real consumers undercut efforts at rational planning, encouraging patchwork improvisations. The very dominance of the capitalist factions in polyclass parties facilitates negotiated settlements with expropriated enterprises: there are seldom ruptures or noncompensatory actions. The extended links between metropolitan and peripheral firms outside the enclave determine the need for moderate settlements; the central issue for the national capitalist class is not to lose the confidence of their foreign collaborators. The policy of mutually satisfactory nationalization settlements, then, becomes symbolic of the larger ties that bind the metropolitan and peripheral firms.

The negotiated settlement from the top and the adequate and prompt compensation require the immobilization of the population. The electoral road defines political activity strictly in terms of the election of the officeholders: the latter then resolve the issues. In the case of nationalization, the negotiated nature of the expropriation and the subsequent appropriation of the surplus by the capitalist class are facilitated by the absence of a massive mobilized popular presence. In both cases, in the favorable arrangement with the expropriated classes and the exclusion of the popular classes, the electoral process allows the national propertied classes the greatest degree of freedom.

Conversely, the nonelectoral road to power, which pushes the populace to the forefront, makes it difficult to sell favorable settlements with the foreign bourgeoisie and creates increased pressure to dispose of the national surplus in directions favored by the populace. The alternative of an elite-directed nonelectoral road to nationalization usually favors the state bureaucratic sectors, minimizing private capitalist pressure but also marginalizing the masses from decisions affecting the disposition of the surplus.

Social Polarization, Nationalization, and Surplus Appropriation

The general formula that allows for private capitalist appropriation of the national surplus accruing from expropriation includes a polarization with specific features: broad alliance of classes, which usually masks the hegemony of the bourgeoisie; low intensity of commitment; and diffuse definitions of the specific goals to which nationlization is directed. The terms of the polarization favorable to the bourgeoisie are to amalgamate conflicting national classes within a national consensus that develops on the basis of a national dialogue and is set in terms of patriotic rhetoric. The central point of this type of polarization is to limit the nationalization to discrete targets without providing new organizational or ideological instrumentalities that could contest the issues of surplus disposition in the postnationalization period. Both the patriotic rhetoric of the private bourgeoisie and the developmental phrases of the state bureaucratic sectors are designed to mystify the essential class issue of who disposes of the surplus.

The central political task for the political representative of the bourgeoisie within an immanent process of nationalization is to define the problem as one of an external enemy and to deemphasize the functional similarities of internal classes (exploitation) and their structural linkages. Conversely, within a social revolutionary process, the appearance of a bloc of classes disguises the hegemonic revolutionary class (whatever its explicit ideology), which eventually extends the nationalization process beyond the enclave to embrace the national capitalist class. In this case, the polarization is prolonged and changes in substance, increasingly dividing the national classes and transforming the national struggle into a class struggle. The initial broad alliance dissolves, class feelings rise, and politicoeconomic goals become historically specific. In brief, the more profound the combined polarization (nation-class), the more likely the disposition of the surplus will favor the masses. The more diffuse the polarization and the less coherent the movement, the more likely the surplus will be used to strengthen private capitalist expansion.

Capital Linkages and Surplus Disposition

When national capital becomes entwined with the foreign enclave or the sector of foreign capital designated for expropriation, the likelihood is that national private capital will resist nationalization. And if nationalization subsequently occurs, it will not be in a position to take hold (at least immediately) of the instruments of state in order to extract the surplus. On the other hand, the separation of enclaves of foreign capital and segments of national capital facilitates the absorption of the former by the latter in the process of nationalization. The degree of integration between national and foreign capital, then, affects the subsequent disposition of the surplus. The greater the integration, the less likely that the national capitalist class can benefit from nationalization.

The integration of national capital on a global basis does not affect this basic proposition since what is at issue is the particular sector of foreign capital that is expropriated. In the long run, however, even sectors of national capital that possessed links with foreign capital can easily recuperate their position and exercise influence over the disposition of the surplus. Likewise, state enterprises that begin in opposition to private capital, following market principles, can easily expand the areas of private activity. The infiltration of individuals from the private sector into the state apparatus and the influence that market relations have on units within the state sector result in the loosening of the controls by the state over economic activity. The state increasingly abandons the more lucrative areas to private activity and assumes the social costs of private accumulation.

Summary

The growth of state capitalism or nationalization of key profitable enterprises can be seen as providing key inputs for the development of private national capital under specified conditions. The conditions include a well-defined and organizationally strong entrepreneurial class tied to the rising political party capable of converting its hegemonic position into an instrument for surplus appropriation. Furthermore, the presence of an influential private capitalist class facilitates the return of foreign private firms as partners (mixed enterprises), as decision makers (management contracts), or as growth managers (technology). Under the pressure of nationalist politics, the pattern of undisputed direct ownership becomes increasingly obsolete. The direct influence of the foreign capitalists declines. The immediate beneficiaries of their decline are those sectors of the national capitalist class most closely tied to the hegemonic nationalist party. And it is precisely these factions of national capital that are in a position to limit effectively nationalization to the point of reintroducing foreign capital (mixed companies) and to redistribute the surplus. The pivotal role of these politically influential factions of the bourgeoisie allows them to become involved in the most lucrative aspects of the new economic plans, since, in most cases, these very same factions design the plans and head the funding agencies.

The nationalization process begins with a declaration of profound and sweeping changes and is gradually unhinged; first, by the constraints imposed by national private capital and later by the accords with foreign capital. According to this sequence, the process of nationalization is largely a means of capitalizing national capital and of redefining the relationships between state and foreign capital. The overall perspective is one of the growth of capitalism with all of its features: concentration of wealth, proletarianization of labor (especially rural labor), and a large reserve army of unemployed. The role of the state is as a provider of capital and arbiter between the competing factions of the bourgeoisie. Increasingly, political conflicts become class-based, including state, national, and mixed enterprises. The nationalist vocabulary increasingly becomes a weapon to discipline the labor force and to heighten exploitation ("to strike against national firms is unpatriotic"). Nationalization, emasculated of its popular content, becomes the historic turning point for national capitalist exploitation.

As was pointed out above, both the form and content of nationalization vary from one regime to another. Thus, it is imperative to distinguish between a nationalization that contributes to the growth of capitalism in one form or another and a nationalization that provides the instrumentality for collective planning of the national economy. Clive Jenkins, in describing the nationalization process and the structure of the nationalized firms in England, posed the essential question: nationalization *for whom?*[1] While some knee-jerk conservatives (in the United States this includes most business and political leaders) continue to flail away at any notion of government ownership, in most countries it is assumed by both conservatives and businessmen that some forms of government ownership in certain areas of the economy are necessary for the development and expansion of private enterprise. Hence, outside the extreme privatism of the United States, public ownership in itself is neither novel nor innovative. What is crucial is determining the role that the public firms play in relation to the private sector and determining who controls and benefits from nationalized property.

STATE CAPITALISM

An Historic Comparative Perspective

State capitalism is defined as a social system in which the principal sources of surplus production are owned and directed by the state and in which the state becomes the principal source of capital accumulation within a market economy. The mere intervention of the state in economic activity is characteristic of other regimes as well as the state capitalist. The mere nationalization of enterprises and/or industries also characterizes other types of regimes: nationalizations have occurred to provide cheap services to private enterprise or have taken place with regard to unprofitable firms. In neither case does nationalization, per se, provide evidence of state capitalism, which requires that the state become the holder of profitable firms that serve to finance the reproduction of capital. Large-scale profitable state enterprises differ from bureaucratic collectivist firms in that the latter are subordinated to the central plan, whereas the former function within the market. The more obvious difference with a socialist form of collectivism is found in the latter's control and direction by the workers and employees involved in production. Hence four forms of nationalization can be distinguished:

Private capitalist nationalization: In the context of a predominantly private capitalist economy, nationalization takes place to bail out sectors of the capitalist class involved in unprofitable but socially or economically necessary types of economic activity, thus enabling the compensated capitalists to invest in newer and more profitable lines of production. The state runs the firms at a deficit, thus subsidizing the costs of production for those private firms using the state sector (transport, communications, basic industries, declining industries, and so on). This experience was or is commonplace in Europe throughout the twentieth century.

State capitalist nationalization results in state ownership of profitable enterprises that serve to capture the surplus for the financing of state and national private investment.[2] The expropriation of raw materials, banks, insurance companies, and basic industries provides substantial sources of new capital toward strengthening the internal (private and state) forces of accumulation. This type of nationalization usually occurs in the Third World and usually affects foreign-owned enterprises.

Bureaucratic collectivist nationalization occurs in the context of total transfer of the principal means of production, foreign and national, to the state within a bureaucratically controlled and planned economy and serves to finance the expansion of national state enterprises. This has been the experience in Eastern Europe and the USSR.

Socialist nationalization of the means of production is part of a total transformation in which the direction and ownership of the process of production are under workers' control. This has appeared in varying forms in Yugoslavia (prior to the advent of the market) and perhaps is present in China.

The state capitalist transfer of power—from foreign capital to bureaucratic capital—is the most susceptible to shifts and/or conversion back to private capitalist production because it does not challenge the basis of market production. Furthermore, the continuance of a private sector opens opportunities for private transactions between capitals (private and state). The historic experiences of nationalization in market economies found in Mexico, Turkey, and Egypt conclusively demonstrate that state entrepreneurship is either a transitional role, which figures in large-scale transfers of capital and resources to the private sector, or else merely establishes the resources, power, transport, and basic industries for national private capital development.[3]

State capitalist expansion has been fraught with contradictions that have limited its potential for sustained and rapid growth and, not infrequently, contribute to stagnation. Incorporated into the political patronage system that characterizes bureaucratic life, the state-owned firms absorb personnel independently of their productive capacity. This results, in many cases, in administrative overhead costs exceeding income, leading to large deficits that are financed by more and more oppressive taxation on the labor force and forcing an alliance of the right and the masses. State-owned firms are subject to external market conditions: large multinational firms control crucial inputs, including technology and know-how, while international banks exact onerous political and economic terms to finance expansion. Public enterprises may compete with one another and leading sectors may deplete less fortunate competitors, leaving an insecure base for long-term development. The results in the above-mentioned countries are known: a return to liberal market economics, the lifting of trade controls, the privatization of industries, and openings for foreign capital. In Turkey, as well as Egypt, statism was the first stage of modern capitalist development, laying the basis for associated capital development projects increasingly involving national and foreign private capital.

While there are important and deep-seated conflicts during the nationalization process between public and private enterprises, the long-term limitations built into state capitalist development provide the basis for an historic convergence of interest. At the level of the individuals involved in the state sector, a number of opportunities appear for private transactions. State subcontracting, credit and loan manipulation, purchases of inputs, merchandising of outputs, and consulting may involve state "entrepreneurs" in activities with private profit-taking firms. Not infrequently, state sector executive decisions promote private investors with whom they later join as senior partners, thus converting a public career into a trampoline for private enrichment and encouraging the spread of private ownership. This conversion from public sector functionary to the board of directors of a private firm is underwritten by the transfer of scarce public resources to the private sector, which grows while the state sector vegetates.

Venezuelan State Capitalist Development

In Venezuela, state capitalist nationalization, which captures the surplus from foreign capital, invests only a part in state-directed and -controlled firms. The rest is channeled through banking institutions toward the private sector and, in some cases, toward new foreign investors:

Nationalized foreign firms (natural resources)	Surplus flows	State	State Private national Foreign firms	Basic industry (upstream) Manufacturing (downstream)

In the short and medium run, state capitalist nationalization is a vital instrument for strengthening national capital. In addition to providing a rich source of new capital funds, its control over allocations serves to promote a whole series of related industries and their offspring. According to CORDIPLAN, the investment strategy involves "selective foreign investment participation" and, more important, "promoting national investment, encouraging public and private savings, as a means of offsetting the effects of the restrictions on foreign capital, channeling it especially toward the dynamic and strategic sectors of the economy."[4] In a 1975 OAS report, the linkages between state and private sectors were made more explicit:

> Intervention by the state as an entrepreneur in the petroleum, gas, and iron and steel industries and in certain special services provides for a broad field for action by private enterprise.... In large measure, public activity has been and will in the future, be directed toward the promotion of priority activities by the private sector. This governmental activity includes the creation of basic infrastructure (as well as monetary, credit, fiscal or institutional measures)....[5]

The nationalization of oil provides the basis for a series of petrochemical plants, followed by plastics, fertilizers, and derivatives. The concentration of wealth in the national state provides the basis for the proliferation and diversification of industry, opening opportunities for state and private capital. The political and economic functions of state capitalist development in the case of Venezuela have been to strengthen the national private as well as bureaucratic capitalist classes. The funding and elaboration of industrial expansion is accompanied by a package of regulations, decrees, and laws whose fundamental intent is to allow the national private sector to capture and control a substantial part of the new industrialization in order to be able to associate with foreign capital on an equal basis.

The new rules of the game for foreign participation can be summarized as follows: "Foreign investment will be welcome, mainly as part of technology packages in which minority equity holding by the foreign investor would be encouraged";[6] local content requirements will be pressed to promote national industrialization; sectors reserved for national companies are specified, including domestic marketing, public services, and communication; a 5 percent reinvestment limitation for foreign-owned firms (except if the investment occurs outside the Caracas area, as part of a decentralization scheme); restructuring in royalty payments between subsidiaries more than 49 percent foreign-held and parent companies;[7] a 15 year fade-out period of foreign-owned firms (100 percent); annual net dividends limited to 14 percent of authorized and registered capital; and equity holdings of foreign banks in financial houses limited to 20 percent and minimum capital requirements increased.

While these measures enhance the position of national capital, there are equally good reasons for foreign capital to participate in Venezuelan development. Corporate taxes are among the lowest in Latin America and are the lowest among the larger Latin American states. Financial operations are far easier than in any other Latin American country because of the well-developed financial system, the ample funds, and the minimal controls on local and foreign financing. Despite some minor limitations in the cost of borrowing, foreign subsidiaries have access to local credit. Furthermore, under U.S. business pressure, Venezuelan officials have modified ("clarified") limitations on foreign capital. The superintendent of foreign investment has allowed that a firm could be considered a local manufacturing firm and not a domestic marketing outfit even if locally it produced less than 51 percent of what it sold; only 30 percent local value added would be sufficient to qualify a product as Venezuelan.[8] The legislation on equity transfers and the limitation on profit remittances and reinvestment were also being considered for modification.

The new rules on the role of foreign investment provide for collaboration between foreign capital and their new allies among the private and bureaucratic bourgeoisie. Foreign investors must share the exploitation, not displace the local bourgeoisie. In summary, the new rules are not very restrictive and compare favorably with those in Mexico and Brazil, long considered lucrative areas of profit-taking activity. The key to Venezuelan state capitalist economic development is to provide more opportunities for Venezuelan investors through state regulation of foreign capital. Thus, regulation of foreign capital and promotion of growth and expansion of national bureaucratic and private capital leads to a new historic bloc of classes in which national and foreign industrial capital collaborate. Within this process, the banking system has a dominant influence on the economy. Through its access to state funds and its linkages to private foreign capital, this system becomes the pivotal area, allocating resources and directing industrialization.

Ties are opening in three areas: state and foreign capital in large-scale petrochemical enterprises, national private and foreign capital in industrialization that extends beyond assembly plants (with both having access to local funds through the private banks), and agrobusiness and foreign capital in marketing and processing of agricultural goods and production of equipment. During the first six months of the Pérez presidency, the nationalist populist phase, when a wave of social and political decrees and laws was enacted favoring national classes and labor, there was uncertainty and therefore reluctance among U.S. investors to join joint exploitative ventures.[9] There was some uncertainty concerning what laws or decrees would be enforced and how far the process would go. Subsequently, the nonimplementation of many measures and the one-shot nature of the wage policies opened up the investment flow, and thus optimal conditions for profitable exploitation were encouraged.

The notion of collaboration between Venezuelan and U.S. capital is further clarified in an interview with President Pérez when he stated:

> Venezuela's position is very clear on the subject. We do not want to bring the economy under complete government control. We believe it is wise not only to have domestic free enterprise involved but also to cooperate and trade with the international private economy. However, we believe that such

relations should be subject to terms that protect our national interest. And we wish to set these standards within a legal system and to guarantee private investment.[10]

He went on to note that, "There is no danger that the behavior of Venezuela will drive foreign investment away ... there is no hostility; we simply set certain conditions that we feel are fair and equitable."[11]

As noted earlier, Venezuela's conditions are far from onerous, since the principal beneficiaries will be private Venezuelan capital, the proposed partners of foreign capital. When the Venezuelan state even provides financial aid to foreign capital, its development approach can hardly be considered antiimperialist, since it promotes imperial growth while strengthening national capital. This is clear in the use of *fondos de credito,* which is a source tapped by foreign capital in forming mixed enterprises.

Venezuela's nationalist rules do not exclude or displace foreign capital, but allow national capital to enter into the picture (financed and protected by the state) before foreign capital, with its superior resources, can preempt investment opportunities, monopolize markets, and so on. The state capitalist formula carries with it, then, an ensemble of state ownership of key resources that, in turn, branches off into a series of mixed foreign and national (private and state) enterprises. Together this package contributes to the growth of capitalism and the expansion of state enterprise.

Statism and the Growth of National Capitalism

State ownership is neither wholly a political nor circumstantial outcome. The struggle over nationalization reflects the crucial importance of key foreign-owned resources for overall economic expansion. In the shadow of the expanding multinational corporations, the nationalist movements continue to exert a constant, if uneven, pressure. Beginning in the 1930s, efforts were made to limit grants of new concessions to foreign companies (measures nullified in 1945–47 and 1956–57), to increase the tax base (50/50 in 1948, followed by 65/35 in 1958), formation of a state oil corporation (1960), effective state intervention in fixing prices (1970), and reversion of oil ownership to the state (1971).[12] This gradual growth of national control set the stage for nationalization.

Over half a century, with occasional reverses, the growth of state control of petroleum occurs concomitant with the growth of an urban petty bourgeoisie that increasingly depends on state allocations and exercises political power but lacks control over economic resources. The constant intrusion of the state and the limitations on foreign prerogatives were, in the first instance, an exercise of power by the propertyless petty bourgeoisie. The parallel growth of urban bureaucratic stratum and state control over its major resource base set the stage for the nationalization.

However, the actual consummation of the nationalization had more to do with the desire and interest of the national capitalist class to harness the nationalization to its own drive for expansion and economic power. For the Venezuelan bourgeoisie in the 1970s, the high oil demand (and profits) and the low risks (of external boycotts, intervention, and so on) provided an excellent opportunity for advancement. Reversion of oil became part of bourgeois consciousness precisely as world market demand made nationalization profitable and OPEC provided guarantees against retribution (interview with Antonio Díaz, Fedecámaras).

The simultaneous growth on a world scale of national forces within economies similarly affected by the petroleum monopolies leads to a worldwide process that coincides in two facets: nationalization and the promotion of national capitalism (in varying degrees). Recent moves toward national control are evidenced in Libya, Iraq, Iran, Algeria, Saudi Arabia, and, to a lesser extent, among the Gulf Oil states. The degree of state ownership varies, as does the type of relationship that evolves with the multinationals. In most, foreign capital, while losing most of the exploitation rights, has managed to hold onto the lucrative marketing, transportation, and refining processes, as well as holding management contracts and selling technology. Nevertheless, the worldwide trend toward national ownership has made it difficult for the multinationals to slide over and increase production at one set of wells to displace or undermine nationalization efforts within any one particular national unit.

Whatever the limitations inherent in any given set of nationalization procedures, from the point of view of a thorough social transformation, the nationalization of production throughout the oil world has provided the first major capitalist challenge to Western dominance since the Third World was colonized. The pivotal fact concerning this state form of nationalization is that the scale of the tasks of accumulation initially developed by the multinationals cannot be assumed by the national private capitalists. By default, the state must assume the role of original accumulation in light of the weakness of the national private bourgeoisie.

State Policy and the Bourgeoisie

To best understand the positive relationship that exists between the Venezuelan business community and the Pérez government, it is important to analyze the government's investment institutions and the manner in which they distribute funds. There are three primary sources for investment financing: the Venezuelan Investment Fund, the Industrial Investment Fund, and the Agricultural Investment Fund. The vast bulk of investment funds in industry and agriculture has been directed toward large and middle-sized private firms. Forty times as much investment funding has been allocated to middle-sized and large enterprises as has been directed toward funding small firms.[13] Almost all funding is through private banks whose links are with big business and who justify loans on the basis of risk and collateral. Obviously, large and middle-sized capitalists have far superior access to the state apparatus than do the small businessmen, President Pérez's rhetoric to the contrary notwithstanding. In 1974–75, the Fondo de Inversiones provided Bs.13 billion for state and big private firms in basic industry; the Industrial Credit Funds provided funds (Bs. 2 billion) for middle-sized and large firms (enterprises capitalized at Bs.6 to Bs.50 million).[14] The Corporation for the Promotion of Small and Medium Industry (CORPOINDUSTRIA) had only Bs.400 million for small and middle-sized firms.[15] If allocation of investment funding is used as an indicator of the nature of the government, the Pérez administration would be considered as representative of the interests of national monopoly capital, with a satellite sector of small and medium-sized firms tailing along.

The allocation of investment resources in agriculture during 1974–75 substantiates the notion that the entrepreneurial class forms the principal base of the current government: 86 percent of loans, amounting to Bs.860 million, went to cattle ranchers and only 14 percent was distributed among farmers, the bulk of whom were large and middle-sized commercial farmers.[16] The enormous disparities led to some protest among the commercial farmers; for 1975–76 the agricultural fund will concentrate loans to noncattle sectors—apparently the commercial farmers outmuscled the ranchers for the moment. These allocations in agriculture suggest a decided shift away from the half-hearted peasant-oriented agrarian reform strategy of the 1960s to a full-scale commitment to large-scale mechanized agriculture. The emphasis in government policy is on enhancing the position of the entrepreneurial sector and the abandonment of the small producers to a marginal position.

The personnel and criteria within the investment fund agencies reflect the current priorities of the AD government. Almost all directors of investment funds are corporate executives and act upon business criteria in distributing funds. Obviously, if one were searching for means to facilitate the smooth flow of resources to the large private landowners, the choice of personnel is ideal. This elaborate tie-in between the investment fund agencies and the large firms in the private sector does not lend itself to a populist interpretation of the regime. Rather, it is a very elitist strategy that has taken efficiency and productivity as its bywords and relegated redistribution of income and rewards to the lowest level of priority.

The internal fragmentation of the bureaucratic structure of the state facilitates private control and the formation of new autonomous agencies closely linked in the name of efficiency, to the private sector. The Venezuelan bureaucracy has been characterized by one official of CORDIPLAN as possessing "multiple centers of decision, atomized and dispersed." Within these circumstances, the criteria for the evaluation of requests for investment funds (in the Ministry of Development, Venezuelan Development Corporation, Agricultural Development Bank, and so on) are based on personal contacts with not much consideration given toward the norms of project design. As a result, planning is haphazard (sectoral projects are not complementary, if not outright incompatible) and entrepreneurial groups who have the personal ties are the beneficiaries over and against small businessmen, peasants, and so on.

One of the key instrumentalities of big business influence is through the control private banks wield over the channeling of funds. The pretext given by AD leaders for abdicating state control in favor of the bankers—and inevitably through them to big business—was the lack of an administrative structure to administer the billions of bolivars designated for investment in agriculture and industry. This argument is nothing but a self-serving rationale that does not deal with the more basic question of why the state machinery has remained divorced from the financial and productive facilities of the country. The reason the state was unequipped to deal with financial allocations was because it was never prepared, nor intended to, by previous AD (and COPEI) administrations, whose main thrust was to develop an economic strategy to accommodate the private sector and a public bureaucracy to accommodate its vast patronage network. It is not surprising, then, that a patronage machine turned state administrator is not able to provide leadership and direction to the development push, but rather is wholly preoccupied with lining pockets (a problem that requires separate consideration).

The Pérez coalition began as a vast amorphous body that included a populist lower class, technocratic intelligentsia with strong nationalist-statist proclivities, an army of petty bourgeois patronage-oriented white collar workers, and high-powered development-oriented entrepreneurs. During the first four months of Pérez's administration the initiative was with the populist masses and the technocratic nationalists, as decrees on wage increases, labor protection, debt annulment, tax concessions, import subsidies, fixed interest rates on time payments and so on were passed along, with the announcement of legislation to nationalize the iron and petroleum industries. As a result of these measures, the popularity of the regime among the masses increased substantially, forcing sectors of the left opposition, the Movement to Socialism (MAS), to support the progressive measures of the government.

On the other hand, these same initial measures of social reform caused dismay among foreign investors and Venezuelan entrepreneurs. Capital was converted into foreign currency and fled the country. The Venezuelan entrepreneur, ever fearful and obsessed by stability, was paralyzed by these initial changes. As a result, beginning in the middle of 1974 and increasingly thereafter, the government moved to consolidate its relationship with the national entrepreneurial class. No new wage increases were decreed, and by the middle of 1975, the 19 percent rate of inflation (May 1974 to May 1975) had wiped out labor's gains for all but the lowest paid category.[17] Prices of food and clothes and shoes were allowed to increase dramatically, while wages were held down, thus reconcentrating income in the hands of the business community.

Speculation on the price of goods was matched by the same phenomenon in land and building. The competitive mechanisms that were to hold down prices were inoperative within a largely monopolistic economy—leading to higher prices and profits—while the government reduced taxes. Thus, while government policy makers were making long-winded speeches about social improvements and government controls and planning, the results of the actual practices of the government were directed toward promoting business opportunities in the hopes that this would lead to greater production. Having chosen the entrepreneurial option, the government did not have any choice but to let salaries and wages lag behind prices, hoping that the added earnings in the entrepreneurial sector would lead to higher industrial investments. Given the resource and income available, the industrial performance of the entrepreneurial group is unimpressive: a 9 percent growth rate. In order to obtain the collaboration of the private sector in its development program, the government was forced to curtail its social welfare program. Thus, while the social costs have been high (in terms of sacrifices in workers' standard of living), the economic benefits are quite sparse (in terms of raising production). Only 15 percent of government funds lent to industry are recovered and government financing has led to the importation of finished goods in high-profit areas. This type of economic activity has led one planning official to describe Venezuelan businessmen as possessing more of a shopkeeper mentality than a true entrepreneurial vocation.[18]

Along with the shift from populism to entrepreneurial promotion, the Pérez government also downplayed the role of the technocratic statists in favor of giving the private sector more influence in managing the planning, programs, and direction of the economy. During the first year, the administrative apparatus was the scene of an obscure but important struggle between the public-sector-oriented technocrats in CORDIPLAN and private entrepreneurial groups over control of investment funds. In the end, despite the fact that CORDIPLAN still retains some leverage, the private sector seems to have won the decisive struggle. The bourgeousie is firmly ensconced in the agencies administering investment funds. It legitimates and strengthens its position through appeals for efficiency and autonomy. The net effect of this is to weaken the position of the technocrats in the state corporations. Within this context, the decision to turn the mammoth nationalized petroleum company into an independent holding company in which inputs and by-products will be in the hands of autonomous firms augurs for a further opening for private sector pillage of public resources.

Thus, in the course of a single year, the Pérez administration, which began with a high-sounding nationalist-populist program, has been converted into a businessman's government that increasingly matches steps with other socially regressive regimes in the region. While accentuating capitalist development, it does so by heightening working class exploitation, maintaining high levels of poverty, and marginalizing the peasantry. While some rightists (the weekly *Resumen*) complain about the excessive role of the state, most businessmen, with their feet on the ground, are reaping the benefits of the oil bonanza, much to the dismay and increasing disenchantment of most of the AD's popular electoral base.

Attitudes and Perceptions of the Entrepreneurial Elite

One apocryphal incident that perhaps illustrates the importance of the private sector within the Pérez administration describes the manner in which the annual plan was formulated. After the initial guidelines were set forth in the ministry, CORDIPLAN, the government planning agency, had to ask Fedecámaras, the private business association, for a copy (the Ministry of Planning submitted the plan for editing and approval to the private association prior to shipping it on to its planning agency). The general optimism found within the business community is well founded. With the passing of Pérez's 90 days of social reform (April-June 1974) and the definition of an entrepreneurial orientation in the subsequent period, the insecurity and uncertainties that initially characterized the business community disappeared. Fiscal spending, favorable wage-price policies, and growing investment funding spurred confidence among capitalists. Central Bank officials openly admitted that the development strategy is geared to big capital, arguing that, for example, in agriculture, loans concentrated in highly capital-intensive enterprises are recoverable and promote import substitutes, for example, meat. In the case of industry, the argument is similar: "the very dynamic of the system leads to concentration of lending to powerful organizations who can absorb the funds and provide the collateral."[19] Thus, funding is directed toward the established entrepreneurs, that is, those with wealth and power.

The acceptance by the government of a major role for the leading private entrepreneurs in the development process has manifested itself in the smooth and accommodating viewpoint that Fedecámaras has taken toward the nationalization process. One entrepreneurial leader describes the planning process as a dialogue between public and private sectors. On the other hand, some CORDIPLAN *técnicos* assert that when one party writes the script and the other reads it, it is something less than a dialogue. The fact that the state has clearly delimited the areas of state ownership to long-term, large-scale projects with a longer maturation period has greatly encouraged private capital, which looks forward to rapid and substantial profits in downstream projects. This division of investment is enhanced by the private sector's influence over the ordering of priorities and influence over the planning mechanism, thus limiting the scope of activity of the statist técnicos by confining them to a subordinate and auxiliary role.

It is in this context that the president of Fedecámaras can voice almost complete agreement with the development approach of the Pérez government, citing the greater incentives for private investment, the broader field for private enterprise, the reorientation of production, and the specification of the rules of the game.[20] The close collaboration between Fedecámaras and the administration is concentrated in planning and in the institutions disposing of investment funds. Within this context it is no wonder that industrial leaders assert that "a strong state does not preclude private participation."[21] Statism is not the issue. The real question is and always has been: state intervention for whom? And in this case the obvious beneficiary stands to be the private business sector.

The leadership of Fedecámaras has, in present circumstances of preeminence within the Pérez regime, developed a conjunctural nationalism tied to support of political democracy. In supporting the government's high price for petroleum, it sees an effective tool in bargaining over the price of industrial imports from the developed capitalist countries. Higher prices in petroleum strengthen the private sector, as indicated by the expansion of private investment. Regarding Andean Pact Decision 24 limiting foreign investment, some entrepreneurs argue that it is a temporary measure to allow national capital to organize a base to later join with foreign capital on a more equal basis and draw on its technology. Decision 24 is, then, a bargaining weapon that the national bourgeoisie uses to avoid the preemption of markets and displacement by foreign firms. The notion of equality between developed and underdeveloped countries is translated into an effort by the national bourgeoisie to gain some footage in its struggle to participate in the capitalist world without losing out in unequal competition.

Nevertheless, there is a persistent gnawing feeling of inadequacy—that in fact the industrial community is just a nation of shopkeepers, businessmen who own assembly plants. This is evidenced by Article Five in the nationalization legislation that allows for the formation of mixed companies. Originally formulated by Fedecámaras in opposition to all major political parties, including AD, the article was later incorporated in the final legislation and supported by AD. When industrial leaders describe the clause allowing mixed companies as providing maximum flexibility, what is usually implied is the fear that if the national bourgeoisie cannot handle production, it wants to have the option of reverting back to dependence on the multinational corporations. It wants to lose its umbilical cord and have it too.

The industrialists' support for political democracy is premised on a clear perception that within current political circumstances there is little or no radical potential inherent in the nationalization process. Recognizing that limited nationalization could be harnessed to national capital expansion, the president of Fedecámaras called for a free play of political forces to allow for diffuse capital interests to intervene in shaping state policy.[22] In this regard, some segments of the Venezuelan national bourgeois are critical of both the Brazilian and Peruvian models. In the former, they perceive the multinationals preempting markets from national capital; in the latter, they note that the state limits the role of private capital. The Venezuelan bourgeoisie, finding readily available state financing for private capital, has begun to discuss and debate plans to diversify its investments. Nationalism, which does not arouse mass passions for social change, is endorsed by the bourgeoisie as a means to achieve growth and competitiveness without confronting the capitalist centers. Capitalist democracy based on political leadership that is "conscious that the motor of development is the private sector" (to paraphrase one industrialist) is supported against "dictatorships which lead to centralized control" and limit the participation of the private national bourgeoisie.[23] Reciprocity, or mutual support, between the state and the entrepreneurial association is based on their common class orientation—the pursuit of a process of development that furthers the goals and interests of the capitalist class. Above all else, the Venezuelan bourgeoisie is heavily dependent on the state to act for it, that is, to take measures to protect, nurture, and expand facilities for private national growth.

The response of the business community to the Pérez entrepreneurial orientation has generally been favorable, with few overt and strident critics. Hans Neumann, president of Corporación Industrial Montana, a holding company for 14 firms, expressed the general sentiment when he stated "I have full confidence in the present government."[24] Fedecámaras reiterated its general support of the Pérez administration soon after the establishment of PETROVEN, the state petroleum holding company, as an autonomous enterprise, declaring its "great confidence" in the country.[25] Fedecámaras's enthusiasm was also manifested in its decision to collaborate in the formulation of the national plan within the framework of joint planning between the state and private sectors.

With a pro-business regime in power, it is logical that its business supporters should call for unity and consensus around its leadership and program; the nationalist rhetoric becomes an ideological weapon to elicit political support for its social program—benefiting big business. Eloy Anzola Montauban, president of a major insurance company, talks of giving a "free hand to the government in the management of industry."[26] In the same vein, Humberto Peñaloza, a corporate executive in the petroleum business, called for confidence and unity in the present government.[27] While supporting the government's initiatives in the area of nationalization because it is the prime beneficiary, Venezuelan big business (despite its newly adopted nationalist posture) is concerned with any eventuality that might adversely affect its other relations with capitalist countries. The same Peñaloza who so ardently defends the nationalization of petroleum and OPEC-proposed commercial agreements with the multinationals to market the oil, partnerships with the multinationals to build a shipping fleet, and other forms of association with foreign capital, qualifies his support of foreign capital by asserting that the ties should only take place on the "international plane" and not the national territory.[28]

The heavy emphasis on concentrating power in the state, while understandable in terms of the favorable socioeconomic disposition of the present regime, is not explainable by it. The Venezuelan bourgeoisie expands as the state expands its facilities. Even in the expansion of industry in the postnationalization period, the industrial leaders call for an active state involvement. Marcelino Barquin, a leading entrepreneur in the metallurgical industry and a director of Fedecámaras, clearly pushed for state-promoted private investment when he noted that "nationalization will not be authentic if we do not construct a metal-mechanical enterprise."[29] He went on to note that he "always supports this type of nationalization.... [its] resources should be a source to promote other sectors of the national economy."[30] Barquin and other like-minded industrialists clearly see the opportunity for private capital in the nationalization action; it becomes a mechanism for diversifying private production and expanding the private sector.

The obsessive concern of the bourgeoisie with mobilizing support around the existing regime and its nationalization program, through calls for national unity, are geared to dampen class cleavages and demands for a share of the new income as well as to provide the state with the freedom to negotiate new ties or associations with foreign capital as the bourgeoisie requires them. The ambiguity of bourgeois nationalism is nowhere clearer than in its demands for national control of international resources as well as its desire to maintain the possibility of forming mixed companies if the external pressures and internal inadequacies demonstrate that the bourgeoisie cannot effectively exploit the national economy. On the one hand, the Venezuelan bourgeoisie favors nationalization; on the other hand, it is overwhelmingly in favor of maintaining "flexibility," meaning keeping the door open to new forms of association with foreign capital. The singular contradiction of the Venezuelan bourgeoisie is its desire for greater independence of action and its need to bring in foreign capital in the form of technology, marketing networks, management, and so on.

The reiterated use of the term "flexible" approach expresses this ambiguous nationalism of the bourgeoisie: its necessity to shift from one direction to the other. For example, the Venezuelan financial bourgeoisie in its *Bank Letter,* asserted that the growing participation of foreign investment in key sectors of production is not always advantageous, citing among the drawbacks the fact that in "many areas foreign capital acts in open competition with local private capital without foreign capital bringing any benefit to the country."[31] The letter went on to blame the foreign use of advertising techniques and methods as leading to unfair competition. Likewise, a study sponsored by the government points to the displacement of local capital in the pharmaceutical industry where 67 percent of the laboratories were foreign owned (and providing no new scientific knowledge in the process).[32]

There is a general fear of displacement by foreign capital within the internal market among the Venezuelan bourgeoisie, which lends itself to support of the state in its efforts at selective nationalization and promotion of industrialization. At the same time, there is a fear of entering into the international competitive arena without the support and direction of the multinational corporations. Venezuelan capital possesses sufficient political leverage to direct the state to promote beneficial economic strategies internally, while it lacks the capacity to innovate and expand internationally, thus expressing its demand for association with imperial capital. The efforts to strike a balance between creating internal opportunities for national capital advance without endangering external ties, markets, and the flow of technology are of prime concern to most Venezuelan capitalists. One of the new directors of PETROVEN, Julio Sosa Rodriguez (owner and director of petrochemical firms, construction companies, and other private endeavors), pointed to the problem of balancing national expansion and foreign dependence in speaking of the need to minimize risks inherent in nationalization without compromising goals.[33] His proposals coincide with those of most other business leaders cited above, calling for specific contracts with multinational corporations for internal expansion and mixed companies for external activity (to "share risks").[34] Sosa correctly points out that the limited nationalization and expanding market situation in Venezuela cement new bonds with imperial capital and inhibits reprisals: "We are recycling through imports alone five billion dollars annually, in addition to the importance of oil to the western hemisphere."[35]

The direction of Venezuelan society is largely determined by an emergent entrepreneurial class. This class supports a selective nationalism and state expansion as measures for sustaining a more equal association with foreign capital. Within the state sector, big business and Pérez have favored autonomous enterprises that facilitate linkages and exchanges with the private sector, lessening the impact of the statist technocrats in CORDIPLAN. Relying on the ideology of productivity and efficiency, the business leaders and the Pérez leadership have maintained the hierarchical structure, salary differentials, and privileges, as well as executive decision-making power, within the nationalized enterprises. The top leadership of the new profitable state sector is composed of private corporate executives, like Sosa, who can be counted on to provide ample opportunity for private sector profits. Fedecámaras, in support of maintaining the previous capitalist structure in the nationalized enterprises, argues that "nationalization means increasing and improving our productive forces."[36]

Having thoroughly penetrated the nationalized enterprises and having locked them into a private capitalist development package, the surplus generated has become the vehicle for a national transformation. Controlled and directed from above and developing new ties with foreign capital, the national capitalist transformation does not provide evidence of a decisive break with imperialism or any serious effort at internal social transformation. This rather incomplete form of national capitalist revolution is probably much more common than the idealized versions of bourgeois democratic revolutions that are described as necessary forerunners to the development of a capitalist society. In the contemporary world it is at least as likely that existing elites will convert to capitalist processes of production as is the possibility that a more democratic social transformation will provide the basis for an expanding market economy. It is within this current of capitalist development from above that the Venezuelan experience must be understood: an effort by the bourgeoisie to lift itself in a more diversified economy through the use of state ownership of resources. Having developed, it hopes to play a major role in exploiting the domestic and regional markets on a par with imperial capital if possible or in association when necessary.

Capitalism and the State

The relationship between state capitalism and the national private bourgeoisie is a complex one that can be looked at from several angles: the role of the state, the economic model, the question of world markets, and the relationship between local bourgeoisie and imperial expansion.

The basic role of the state does not change; in many ways the earlier subsidies and transfers from the public to the private sector are continued, but on a grander scale. But the level and the areas of involvement of the state in the economy may call for a redefinition of the state in its relationship with capital. Both the scale and scope of state operations have created vast opportunities for private expansion on the local and regional levels that heretofore were only very vaguely perceived by the private sector, hidden behind their protective walls and expanding slowly in a limited internal market. The state fulfills several basic conditions for capital expansion: it provides the funds, it creates the markets, it manufactures the intermediary goods, and, through loans, it creates external demand and markets. The state has changed from its dependent status concerned with protection, repression, revenue collection, and providing the minimal services for the maintenance of the established order to an emerging national capitalist political formation capable of creating an internal market, fostering industrialization, and attempting to develop its own dependencies.

The economic model that is emerging—the mixed economy with state ownership (and partnership with foreign capital) in basic industry, mixed firms in intermediary goods production, and private ownership in manufacturing—

is an effort to promote a national capitalist economy. The heavy emphasis on state ownership in the initial phase of accumulation is seen as creating the basis for a multiplication of private sector activities downstream in the more profitable areas of investment. Nationalization and state capitalism become the only means to overcome the inadequate resources and capabilities of strictly private sector activity that had become immersed in a virtually stagnant economy.

The economic model chosen by AD differs in some essentials from both the Peruvian and Brazilian approaches. While all three function within a capitalist framework and all envision roles for state, national, and foreign private capital, there are substantial differences in the mix of each. Brazil is probably the most advanced of the three in the sense that it has gone through a development sequence in which statism was a major early factor, followed subsequently by private national and now foreign capital in association with state capital. Peru is in the statist period, laying down the basis for a liberal national private resurgence. While Venezuela has just entered into a statist period, heavy initial control by the private sector with premature preemption of investment funds may prevent the state from developing the basic industries essential for downstream expansion. The common sequence of capitalist development in Latin America begins with statism (heavy state investment), followed by a liberalization, or destatification, which is usually a prelude to denationalization.

The Pérez regime, in order to accelerate the process of capitalist development in Venezuela, is following a strategy that focuses on limiting foreign capital activities to the external sector (commercialization), fragmenting the operations of the multinationals (buying technology, signing management contracts), limiting exploitation to specified time periods, limiting access to local capital, and directing foreign industry to invest in export activities, as well as selective nationalization, that is, expropriation of particular enterprises by the state in order to fund national capital.

In this national capitalist developmental model, the national bourgeoisie attempts to assert hegemony over foreign capital in order to be able to exploit the labor force and resources. The national bourgeoisie (bureaucratic and/or private) is the agent of capital accumulation, but at the expense of the labor force—reconcentrating capital in its own hands. This type of elite nationalism can shift to either a neocolonial or populist model of development because the private national bourgeoisie is generally numerically and economically in a weak position.[37] The initiatives for this form of capital accumulation usually come from the state, which can impose the conditions on both labor and foreign capital.

Because the state has the central role in maximizing national bourgeois interests, it has two functions, anti-imperialism and disciplining the labor force. In practical terms, the national development state attempts to redefine the terms of dependency to favor the national capitalist stratum and contain labor demands. While selectively squeezing the foreign mineral sector, this approach shares with foreign industrial capital an interest in maximizing exploitation of the labor force: maintaining production, labor discipline, and popular demobilization. The success of this type of establishment nationalism depends on the avoidance of confrontations with foreign sectors and the labor force. Threats from either side may cause the national bourgeoisie to seek alliances—with populists if threatened by the imperial power or with the latter if threatened by the left. While the rise of establishment nationalism may have been influenced initially by radical nationalist pressures, the usual tendency has been for it to dissolve in a series of external agreements that erode the original national development project.

The existing high levels of inequalities of Venezuelan society and their perpetuation by the presidential regime now in power lead to a limited internal market for an expanding Venezuelan capitalism nurtured by state capital. The solution proposed is overseas expansion, especially in the Caribbean, Central America, and the Andean countries. Given the profit margins demanded by the private sector (averaging 40 percent rate of return for each bolivar invested) and the low purchasing power of the vast majority of Venezuelan workers, peasants, and unemployed, the economy could not completely absorb the new funds. Hence, the effort to loan and invest abroad should be understood as an attempt to place seed money for future joint ventures. Under the aegis of state loans, the Venezuelan bourgeoisie is looking for the first time at the possibilities of participating in the world market.

Venezuela's overseas financial transactions are multipurpose: to collect interest on capital that cannot be invested internally to develop overseas economic enterprises that provide raw materials for industrial use, and to provide export and import, as well as investment opportunities, for Venezuelan industrialists.

The general thrust of Venezuelan overseas financial transactions has been toward the international and regional banks and the Caribbean and Central American countries. In 1974, Venezuela lent $500 million to the Inter-American Development Bank; $500 million to the World Bank; $540 million to the International Monetary Fund's oil facility; and about $700 million to the Central American Bank for Economic Integration, the Caribbean Development Bank, the U.S. Emergency Operation, and some Latin American countries. In 1975, Venezuela established a $60 million trust fund in the Andean Development Corporation and a $40 million one to the Central American Bank for Economic Integration.[38]

In addition, Venezuela plans to lend $500 million during 1975–80 to Central American countries and Jamaica to offset the high price of oil and to finance local development projects. By pumping money into the islands and

Central America, Venezuela hopes to find markets and raw materials to spur its industrial firms. Loans to development banks will spur industry, agriculture and agroindustry, and tourism. The most important agreement thus far is the Jamaica-Venezuela bauxite and aluminum for oil deal.[39] Jamaica agreed to provide Venezuela with 200,000 tons of aluminum a year and 400,000 tons of bauxite, the latter building up to 500,000 tons in the fourth year—both arrangements to last for ten years. In return, Jamaica is to pay only half the world price for crude oil imported from Venezuela, the difference regarded as a loan on deferred payment at 8 percent, the money to be used by Jamaica for importing goods, primarily from Venezuela, for its development program. A similar transaction with Guyana involving bauxite and aluminum for oil is anticipated.[40] Financial ties are also in the offing with St. Kitts-Nevis, Grenada, Antigua, and Barbados.

The financial activities of the Venezuelan state have facilitated the movement of Venezuelan private capital. Venezuelan firms have moved into the cement industry of Jamaica and port development in St. Lucia.[41] The entry of Venezuela into the Caribbean has challenged the predominant economic position of Trinidad and Tobago in CARICOM, thus evoking the wrath of Trinidad's prime minister, Eric Williams, who promptly dubbed the Venezuelan efforts a form of recolonization. The external push of Venezuela toward Latin America includes efforts to promote joint enterprises with Brazil and private multinational corporations in the Latin American Economic System in SELA, thus challenging Cuban efforts to promote state-directed firms.[42] The overall effort to find external markets tied to regional pacts is a means of escaping the confines of the internal market without having to face the stiff competition in the international market. Venezuelan capital, as it moves from dependent to associated status, seeks to establish a sphere of influence in the Caribbean and Central America, while sharing markets with Brazil.

The process of recycling oil dollars within the national economy to stimulate rapid, large-scale industrial growth has provided the national capitalist class (private bureaucratic) with a weapon to bargain with foreign capital. In exchange for access to the internal market (and even local financing), the local bourgeoisie hopes to secure technology, new product lines, brand names, and so on. Out of the process of nationalization and the recycling of capital into the national economy and industrial expansion emerges a new association of foreign and domestic capital. Through state capitalism, the national bourgeoisie finds a way to insert itself into the process of imperial expansion. The limitations of the internal market and the resultant constraints on growth (stagnation) forced the bourgeoisie to redefine its relationship with imperialism: expropriation of one sector (oil) in order to extend and deepen its ties with other sectors. A national industrial bourgeoisie with regional aspirations develops through the growth of state capitalism and links up with imperial firms.

NATIONALIZATION AND THE INTERNATIONAL CONJUNCTURE

Politics is not a science because it is the art of acting in unfolding and changing circumstances. The process of nationalization in Venezuela was as much the product of external developments as it was of internal political pressures and economic imperatives. The extremely favorable international conjuncture included a concerted movement of diverse national political forces (OPEC) grouped around a common product (oil) in short supply and essential to imperial growth and expansion. At the same time, the antagonistic imperial consumer forces were badly divided organizationally, competing for advantage and lacking a common strategy. The timing of the confrontation also favored the producing countries: the culmination of a decade of common work and joint membership in a common organization. In addition, the consumer nations had just witnessed the spectacle of a major defeat of the principal core country—the United States in Vietnam—in its efforts at military conquest. Efforts to promote colonial revivalism were confined to the pages of isolated journalists in the United States and the pronunciamientos of U.S. political leaders. Europe was not following. OPEC, by taking a collective decision to raise prices, and the concomitant move by several Arab countries to nationalize their oil holdings, took the heat off Venezuela. Nationalization, because it was part of a worldwide trend, could not be isolated and undermined, as was the case of Mohammed Mossadegh in Iran 20 years earlier. The United States and its Western allies had no levers to pull; to close down markets to one was to cut off supply from all. The consumers had no choice but to look for ways of minimizing losses through compensation for nationalization, management contracts, technology leasing, marketing agreements, and so on. The convergence of a favorable international conjuncture and internal imperatives (political pressure and economic stagnation) produced the Venezuelan nationalization in 1976, an event that, in earlier circumstances, would have been unlikely.

IMPEDIMENTS TO GROWTH AND EQUITY: THE PRIMACY OF THE POLITICAL

The efforts at "sowing the oil" to produce industry have not, thus far, been successful in Venezuela. The major growth has been in services, imports, and commerce. The major obstacles to recycling oil wealth into industrial growth include a capitalist class that has profited from nonindustrial investments, thus dissipating investment loans; a bureaucracy that is incapable of operating a productive sector; and a political system whose terms of success are premised on large-scale funding that, in turn, obligates the winners and their supporters to reimburse themselves through the public treasury—corruption becoming a way of life.

To many observers of the Venezuelan social structure, the behavior of the local capitalists more closely approximates that of a shopkeeper than of the Schumpeterian entrepreneur. Many of the industries are largely assembly plants based on semifinished imports whose design and elaboration have been the product of ingenuity elsewhere. The primary concern of the Venezuelan capitalist is selling and renting, not manufacturing. A large proportion of loans and credits directed toward the industrial sector is funneled by its recipients through commercial outlets, real estate, the importation of consumer items, and finished or semifinished industrial products. Little effort has been directed toward long-term, large-scale projects in basic industry; even less attention has been paid toward fostering innovation, research, or the design of new technology. By demanding and getting government subsidies and protection, extremely high profit margins are maintained, which, in turn, allow the private sector to pay royalties to foreign manufacturers in order to borrow technology and designs or to import semifinished goods. The overhead costs, then, for maintaining the private sector are so high that it is no wonder that the performance of the industrial sector over the last decade has been so dismal in relation to the amount of resources at its disposal.

Alongside the undynamic industrial sector is the state apparatus, long a source of employment for many hundreds of thousands who cannot be productively employed by the private sector. The state, the rich state to which initially accrues the petroleum wealth, absorbs what the private sector cannot employ. And since the private sector is not growing very quickly, the state bureaucracy grows by leaps and bounds. Lacking a dynamic industrial sector that might have given a purpose and direction to state activity, the bureaucracy as employment agency becomes a cumbersome, sprawling labyrinth without capacity to design or execute development projects, except in the area of consumption (housing). Lacking a tradition of productive activity, born in a society that prizes commercial skills and consumption based on ready-made exploitation of petroleum wealth, the capacity of the bureaucracy to engage in productive activity is very low. The gap between plan and execution is abysmal: only 25 percent of the projects scheduled for 1974–75 were executed.[43] Whatever entrepreneurial talent exists in the public sector is quickly recruited by private corporations that, because of their exorbitant profits, can offer higher salaries and perquisites. The state is depleted of its executive-entrepreneurial talent and overloaded with middle and low level functionaries. The entrepreneurial talent recruited to the private sector is, in turn, transformed into highly paid financial-business associates whose principal function is to secure loans and promote speculative enterprises, not productive innovations.

Corruption is part of the social system. In any capitalist society, buying and selling is one of the basic elements of the political as well as economic systems. What is new in Venezuela is the scope and depth of corruption and, in consequence, its impact on development.

Corruption and Private Capital Accumulation

A mass party, oriented toward the entrepreneurial class, offers few possibilities for sweeping structural changes. In agriculture and industry, the AD party has shifted its priorities from redistribution to increasing production and efficiency. Yet, an electoral party that hopes to stay in power must reach out beyond its primary constituency within the capitalist class. The principal instrument through which it stays in power is a vast patronage network that provides opportunities for individual payoffs and rewards to the party loyalists. Large-scale corruption is the necessary outcome for sustaining political loyalties among the lower and middle levels of society when the basic direction of economic policy is directly tied to the promotion of capitalist development. The shift toward an entrepreneurial direction has resulted in corruption substituting itself for social reform as a means of holding together the nonentrepreneurial groups in society.

The scope and scale of corruption are evidenced in the very declarations of AD political leaders, who, fearful that the corruption will erode the party's social base, attempt to exercise some restraint. The decline of populist ideology and practice is matched by an increase of corruption and bureaucratic controls. None other than ex-President Betancourt was forced to denounce the corruption that he felt was weakening the party's support, threatening to "kick out the window those comrades in the public administration who are caught in the act of stealing."[44] Condemning the party functionaries who were ostentatiously displaying the prizes of corrupt practices (buying the latest Impala and staying at the best hotels), Betancourt called for a moral regeneration of the party—without providing any new social or political programs or ideas to motivate it. This is understandable because Betancourt shares the developmentalist orientation of the current administration; his major preoccupation is that the promotion of business be done by honest functionaries. Betancourt also perceptibly took note of the bureaucratization of the party and called for the formation of a new school for cadre to be trained in party doctrine. But neither Betancourt's denunciations of corruption, his shrill and obsessive denunciation of the left, nor his organizational proposals will perform the task of remotivating party functionaries when the party is so deeply involved in the pursuit and promotion of private gain.

Much more perceptive as to the source of corruption within the AD administration is AD trade union leader José Vargas, president of the Venezuelan Workers' Confederation, who noted that

> the statified firm cannot remain pure in a corrupt society. Capitalist society is the generator of corruption: the struggle to secure markets, within the framework of a philosophy of free enterprise, is the source of corruption. A businessman is successful in an entrepreneurial organization in the measure in which, breaking moral standards, he ruins his competitors. . . . I affirm that general corruption is a product of social organization.[45]

And it is that form of social organization (capitalism) that Vargas' party is actively promoting through its policies and practices.

Despite the admission of the secretary-general of AD, Luis Pinerua, that under his party's governance "corruption has become scandalously pervasive in our country," the response—the formation of an Ethics Commission within the party—is woefully inadequate.[46] Efforts to provide limits on contributions and to check political payoffs will be evaded; and moral exhortations and promises to eradicate corruption will fall on deaf ears. Secretary-General Pinerua's statement that the party ought not to accept electoral contributions from businessmen receiving government contracts is a clear admission that these types of transactions do in fact characterize the political process. Opposition Congressman Leopoldo Ferrer has pointed out the nature and consequences of this type of transaction between governing party and private enterprise:

> When a private business or person contributes economically to a party, he feels he has a right to demand favors from this party when it arrives in power. At the same time, the party that receives the contribution is committed to this private business or person and consequently feels the need to fulfill its obligation when it arrives in the government.[47]

With the growth of state resources, corruption has grown as more and more enterprises perceive opportunities to expand and can do so only through the expenditures and financing of the state. At the same time, the growing opportunities for international investment (as well as deposits in foreign banks) make it more difficult to apprehend the administrative officials. More important, the line between a government decision promoting business interests legally through concessions, loans, and subsidies and government officials promoting the same interests through "illegal" means is a very difficult one to draw. Since business is the beneficiary in either case, administrators may feel no qualms in raking off 10 percent for their promotional efforts. Given the crucial importance of the state, as financier, customer, and supplier to private enterprise, and given the administrative overlap and general incapacity, corruption becomes a means of facilitating the flow of government resources to promote capitalist development. From the capitalist development perspective it is only when corruption is so pervasive that it siphons more funds away from expansion that it becomes a problem. And it is precisely this danger that caused Betancourt and other AD functionaries to denounce the excesses.

One further aspect is worth noting—the tendency for high officials with low initial capital to enrich themselves through the public treasury in order to move into large-scale production. In this sense, corruption is a form of original accumulation, a type of primary capital formation that perhaps is typical of all capitalist societies. As Honoré de Balzac once noted, "A crime is usually behind every great fortune." The growth of large-scale cattle ranching is a good indication of the mix of legal and illegal activity. The vast bulk of government credits was distributed to the cattle barons, ostensibly to promote entrepreneurial farmers and lower imports of food. Some credits were also used as COPEI leader José Curiel observed, as a means of paying back electoral campaign contributors.[48] The corruption issue, then, is subordinated to the larger issue of the class nature of the development policy chosen by the administration. Drawing support from the entrepreneurial groups on the basis of promotion of their interests, it is naive to believe that they will merely contribute their meager votes to the campaign. The only resource that the economic elites possess to counterbalance their small numbers is financial power—the capacity to influence and manipulate "popularity." The successful use of financial resources is one reason why Venezuela maintains a highly stratified society within a parliamentary system.

The degree to which AD has raided the state treasury in its quest for individual wealth and esteem has forced even its national leaders to denounce publicly the party's unrestrained behavior. In large part, these leaders sense that the new wealth and widespread corruption will drive an irreparable breach between party functionaries and their electoral base. Corruption has had a double role in this contest: it pays off low level followers who otherwise would receive no social benefits from AD policies (as members of the lower salaried or wage groups) and it provides "working capital" for the upper and middle levels of the party who aspire to entrepreneurial roles and/or upper middle income lifestyles. The exhortations against petty corruption by the party leaders fall on deaf ears as long as the party leaders enter into large-scale financial transactions that are in part provided for by state funds. The populist elan that characterized AD during the 1940s and 1950s and the reformist ideology of the 1960s have been replaced in the 1970s by a decided commitment to the rhetoric of entrepreneurship. The ideological commitments to social reform have been so loosened and the promotion of individual mobility so widespread that it is little wonder that AD party members inside and outside the government have joined in the "dance of the billions," the channeling of public funds into private pockets.

Corruption is, then, a cause and consequence of a consumer society in which private enterprise substitutes commercial and speculative activity for productive investment. The massive influx of oil wealth has merely heightened the degree to which political pay offs shape the behavior of already patronage-oriented party elites. Possessing large sums of money, they have moved to expand their real estate and other property holdings. The channeling of public funds into private gain has seriously impaired the regime's capacity to define national economic projects and, in turn, the lack of industrial dynamism provides a ready excuse for private speculation. The overall result is a spectacle in which AD trade union leaders engage in bitter struggles, not over the erosion of their members' standard of living but over their own admission in the Jockey Club. Likewise, AD youth leaders, who travel to and from provincial meetings in the latest Impala and are lodged in the most expensive hotels, are not likely to have their exhortations for "work and sacrifice for a free and sovereign Venezuela" taken seriously by the unemployed in the ranchos and by the poorly paid operatives in the factories.

In a political system in which the public treasury contains so many offerings, it is hardly surprising that the elections that provide the opportunity to collect those offerings are a central factor contributing to corruption in Venezuelan political life. The length and scope of electioneering (in which more money is spent per capita than in the United States over a two-year period) create enormous political debts that the winners are obliged to pay after the elections. If the wealth accruing to the public treasury after the oil price increases is added to the declining social commitment of the AD leadership, and the generalized patronage and payoff mentality of the middle and lower level functionary party activists is considered, there is some basis for accounting for the profound depths of corruption that afflicts the Venezuelan polity in the 1970s. Where political corruption erodes the very basis of any long-term sustained development effort, AD cannot for long hold popular allegiances—hence the call by Betancourt for "cleaning out corruption" and repressing AD's leftist competitors. Betancourt's political experience tells him that the path chosen by his party will ultimately contribute to a decline in its popular support. Unable to change the party's corruption he so vehemently denounces, because he favors the social and economic policies that promote it, he counsels the repression that will perhaps prevent his competitors from reaping the benefits of his party's failures.

The Impact of the Entrepreneurial Approach on the Class Structure

It is too early to give a thorough account of the social impact of the entrepreneurial strategy of the Pérez regime. Nevertheless, there is a logic to the process and there are indicators that foreshadow the future configuration of social forces.

The drive to promote heavy industry and to provide competitive exports has yielded a development plan that banks heavily on up-to-date technology. As a result, it can be expected that industrialization will have only a minimal effect on the unemployment problem. In this regard, the proposals by Meir Merhav promoting a parallel economy and Decree 21 of the Pérez government providing make-work (encumbering automatic elevators with operators) are merely increasing forms of disguised unemployment.[49] This type of employment provides neither skills nor adequate remuneration. More important, it does not substantially reduce the unemployment rate over the long run.

The second element in the entrepreneurial formula is the loosening of price controls, restricting wage increases, and decreasing social welfare benefits —thus increasing profits and other incentives to private investors, while depressing the standard of living of wage laborers. The conservative daily, *El Universal,* noted at the end of the first year of the Pérez government that "... in general terms the quality of life has not improved substantially, owing to the inflationary impact on the real purchasing power of the economically weakest levels of the population (70 percent of 12 million inhabitants).... Notwithstanding the general increase in salaries and wages of 15 percent, the rate of [price] increase was equal...."[50]

During 1975, there was a deterioration of the standard of living of the working class. The Jesuit economist, Manuel Pernaud, pointed out that the price increases of basic necessities exceeded wage increases. He cited as an example sugar, which "costs 1.5 bolivars today [March 31, 1975] when it was valued at one bolivar [a year earlier],"[51] though he could have gone on to consider increases of 25 percent in food prices as well as substantial rises in clothing and shoes, which exceed wage increases.[52] As with the industrialization strategy, the price policy is designed to promote capital accumulation of the industrial and commercial classes at the expense of the working class. This policy is promoted by the regime, even though it is a well-known fact within the various government agencies that higher profits do not necessarily lead to increasing investments in productive activity.

In housing, the Pérez regime built less than half the number of houses in 1974 (31,405) as were constructed in 1973 (70,000).[53] What is more significant from a class point of view, the public sector, Banco Obrero, which builds for low income families, constructed less than the private sector. In education, the budget was substantially increased to Bs5 billion (over $1 billion) and yet its impact on the quality of education was minimal because, in large part, there was no serious effort made at designing and implementing an overall plan. The lack of commitment in a social policy oriented toward improving the conditions of the working class is the other side of the coin of an economic policy that provides a profusion of resources for the entrepreneurial class.

The third element of entrepreneurial approach involves the intensification of exploitation. This can be seen, for example, in the systematic nonenforcement of the health and safety codes in factories. According to a government study by the Office of Inspection and Evaluation of Professional Risks and Occupational Medicine, 38 percent of the workers employed in a sample of 28 firms (792 workers) were afflicted with occupational diseases.[54] The industrialists' single-minded obsession with profits has resulted in frequent warnings from the Ministry of Health, but rarely have the threats been enforced. Both the regime and entrepreneurs have no intention of lessening the rate of exploitation, since both envision the unhindered expansion of capital as their primary goal.

The fourth element in the entrepreneurial formula is popular repression. To contain the protest of the working class and peasantry, the government has initiated a series of repressive measures, selectively applied, which confine politics and social struggle to the parliamentary-electoral arena, where the government party predominates. In Miranda, the repression of peasant organizations forced the moderate COPEI parliamentary representative, Pedro Humberto Calderón, to protest the government's harassment and jailing of 300 peasants, while landowners occupied public lands without the intervention of the National Guard.[55]

There is little doubt that the present level of government repression could increase substantially if the opposition begins to mobilize effectively the increasingly discontented electorate that is abandoning the AD. The clearest indication of this growing discontent was present at a meeting of middle level AD activists who almost uniformly attacked the government's failure to implement its minimum promises to the lower class dwellers of the barrios. Luis Molina, AD barrio leader, pointed to the contrast between the electoral promises and the party's performance in government: "Here is the party, here are the Carlos Andrés Pérez kids who went to the barrios making promises ... and now every time we go to a barrio or to a hollow we are asked, 'Where are the results?' The governing class must realize that if it's true that we believe in President Carlos Andrés Pérez, the people are waiting for the works that were promised."[56]

The growth of arbitrariness, arrogance, and repression—mixed with anti-Communist demagogy—is directly related to the failure of the regime to comply with its electoral promises. COPEI parliamentary leader Eduardo Fernández cited a survey in which over half of the Pérez voters expressed discontent and frustration with his regime.[57] He went on to argue that the unfulfilled program and the arrogant style of rule were "a grave danger for democratic stability."[58] President Pérez apparently is aware of the fragility of the social basis of his new entrepreneurial development program, pointing out that his regime was the last opportunity for democracy.

This points to the fifth and final element in the entrepreneurial formula: a loyal, reliable military and police force. The possibility exists that with Pérez' commitment to nationalist capitalist development and the resultant erosion of AD's electoral base, a tough-minded civilian or military regime may be necessary to sustain the effort and contain the popular discontent through authoritarian, nondemocratic methods, similar to those of the Betancourt-Leoni regimes of the 1960s. There is little doubt that Betancourt's preoccupation with MAS influence in the armed forces is premised on his understanding that the military must be present as an effective instrument in AD's policy of repressing popular discontent and harassing opposition trade union and party activity. The National Guard and the police forces will become important levers in latter stages of the development experience. Hence, AD seeks a monopoly over their ideological training and political manipulation. Any sharp downturn in oil prices or production resulting in severe cutbacks could lead to a breakdown of the coalition of bankers, businessmen, military, public sector employees, and the trade union apparatus. Under those circumstances, the military would step in—with either a Brazilian or Peruvian style of development.[59]

The social impact of past Venezuelan development strategies can provide us with clues as to directions that the current entrepreneurial approach will take. The long-term effects of this approach to development are most visible in terms of the food intake and health levels of the population. A study presented at the Fourth Latin American Congress on Nutrition and Diet revealed that 50 percent of Venezuelan youth between 1 and 15 years exhibited problems of malnutrition.[60] Among those who die because of vitamin deficiency, 70 percent are less than 5 years old.[61] The National Institute of Nutrition found that, among children age 1 to 6, 55 percent were suffering from malnutrition, while 54 percent of those between 7 and 14 years of age were in the same condition.[62] Over 80 percent of the deaths attributed to youngsters under 5 years of age were due to malnutrition.[63]

Pérez' accentuation of the entrepreneurial approach embraced by his predecessors promises to exacerbate long-standing social problems. The so-called solution to the problem proposed by Pérez' Health and Social Welfare Minister, Antonio Parra León, involves educating individuals and families to proper eating habits.[64] Obviously, the lack of income, high prices, and other such reference points for good eating habits are absent from the minister's sight —and necessarily so. Because the regime is promoting the interests of the investor class, the Pérez government will not redistribute income or lower prices to provide the resources for an adequate food intake. Hence, the government's policy is to blame the victims: their malnutrition is a product of bad eating habits, just as unemployment is a problem of the "unskilled labor force."[65] Needless to say, China and Cuba, both with large unskilled and undereducated labor forces, were able to increase caloric intake and provide full employment, but then their development models did not have to concern themselves with the profit margins of the private sector—a vital necessity for the Pérez regime.

With a rising rate of inflation, a slowing down of the growth rate, declining petroleum income, drops in capital expenditure, and increases in current expenditures, it is clear that the development model is facing serious problems. Oil still provides 75 percent of the budget income, despite the rhetoric of "sowing" the wealth. The promotion of private capitalist development has led to wholesale corruption (under the budget heading "current expenses") and increasing prices and profits, while production inches forward and the standard of living of the populace declines. While the cost of agricultural goods increases, production lags. Industry expands but unemployment remains high. Decrees are passed but not implemented, buried in a corrupt, cumbersome bureaucracy honeycombed with careerist party functionaries whose prime purpose is to line their pockets.

The entrepreneurial approach clearly projects a sharply differentiated class structure in which income, education, and health services are increasingly unequally distributed and in which the state increasingly serves the upper classes, leaving the masses to their private devices, occasionally cajoled with electoral gifts or, when they get out of hand, repressed by the National Guard or police. The regime's development project clearly provides broad opportunities for entrepreneurial expansion and the proliferation of opportunities for the professional class with a business vocation. The policy implications of this approach lend themselves to the notion that income is being reconcentrated in the hands of the upper and upper-middle strata with smaller portions directed toward the propertied petty bourgeoisie. The class composition of the regime's beneficiaries then begins at the very top of the national stratification pyramid and permeates toward the middle, excluding the bulk of the urban and rural labor force.

How effective this approach will be in neutralizing the predominantly petty bourgeois character of political leadership remains to be seen. To the degree that the opposition is enveloped in petty bourgeois milieus—such as the university, the Parliament, the professions—its capacity to grow will be handicapped by the government's selective expansion that can absorb the bulk of the constituents from these areas. Nevertheless, the uneven and unequal growth of income and government allocations is transparent and is creating a readily available mass of supporters whose political and social weight could serve to energize new class-anchored political and social movements. There is a gradual shift away from the old political and social alliances between overlapping classes and strata of business, professionals, workers, and farmers toward class alignments in which small farmers and recently landless workers are increasingly in conflict with commercial capitalist farmers. Urban workers struggle against national private and state capitalists. The old nationalist-populist coalitions against the traditional oligarchy and foreign capital have become less important.

Nevertheless, the new class structure centered on the conflict between national capital and wage labor in the cities and the countryside is far from a replica of the class structure in the imperial countries. There is still a vast army of public employees, at least twice the size of any normal developing country, and a huge reserve army of unemployed, as large as that of a country in permanent depression. The bourgeois-working class conflicts will be complicated by efforts to win over these numerically powerful strata who are neither in the old system of power not yet incorporated in the modern factory system.

The Left Opposition: MAS, MIR, and MEP

The Venezuelan left today is made up of three major political parties and scores of isolated, ineffectual sects and publications. If the Pérez regime and the AD-COPEI biparty system that has predominated over the last 18 years is to be challenged from the left, it will come from among the three major groupings. These three groups have different political origins and prospects. The MIR and the MAS have their roots in the extraparliamentary struggles of the 1950s and especially 1960s where some of their leaders and supporters were active in the guerrilla movement, shifting tactical gears in the late 1960s toward mass struggles. The MEP began, in contrast, as a splinter of the ruling AD party, largely composed of dissident parliamentarians and government officials who felt they were not being taken into account by the party bosses.

Combined, these three leftist forces could provide a meaningful platform from which to challenge seriously the dominance of the AD apparatus over the working class. In competition, they tend to weaken their forces and fragment their impact. Either a joining of forces or the clear eclipsing of one of these contenders is necessary for the left to grow into an organizationally and politically coherent force.

The MEP probably is the largest of the left forces, both in terms of electoral support and representation in the trade union movement and in civic associations.[66] Its ties with the trade unions persist and, in areas like Maracaibo, it holds the leadership of the petroleum workers union and is influential in several other state unions. Basically, built as an offshoot of AD and maintaining many of the political methods of the parent party, the MEP has lacked sufficient leverage in government to provide the favors and patronage to sustain a part of its electoral following and supporters. As a result, there has been a falling off in two directions. Older activists have drifted back to the AD where the patronage action is located, while younger members and trade union militants have joined the newer, more aggressive MIR and MAS organizations. Caught between its parliamentary nationalist-populist past and the present-day competition from the new socialist parties, the MEP vacillates between a radical socialist position, which antagonizes the older professionals, and a parliamentarian activism, which isolates it from the lower classes. Clearly, the MEP has yet to adjust to its new role as a socialist opposition. Unless MEP develops a new type of organization that develops organic ties to the ongoing struggles of workers between elections, it will be outdistanced by its socialist competitors. Its political decline can be arrested by the introduction of new leadership or its fusion with one of the other socialist parties.

The MAS is probably the best known leftist opposition group in Venezuela[67]—and with good reason: it contains most of the leading intellectual and artistic figures on the Venezuelan scene. As a matter of fact, its main strength initially rested with its intellectual, professional, and university followers who served to bring the MAS into national prominence through their propaganda. The attractiveness of the MAS is to be found in its programmatic appeal toward an open Marxism, free from ideological domination by established power centers, and its call for an understanding of specific Venezuelan realities as a basis for developing any revolutionary strategy. Its effort to overcome the sterile and sectarian squabbles that characterized Venezuelan left politics and to reach out to the great mass of politically uninitiated Venezuelans is laudable and promising. Its careful analysis of the existing regime and its efforts to specify the conditions and possibilities within each conjuncture are indeed signs of mature realism that augur well for the movement. More important, its decision to work within the working class movement and to develop strategies and tactics within the interstices of the legal-political structures, as opposed to the rural guerrilla strategy of the 1960s, is a policy option that demographic, political, and economic realities made imperative. It is not surprising that the MAS (along with the MIR) has been one of the fastest growing movements on the left, having grown sufficiently to establish a national presence, though obviously far short yet of making any serious bid for power.

With regard to the Pérez regime, the MAS has followed a line that has varied with regime policies but essentially moved from critical support to constructive opposition. During the first five months of the Pérez regime, the MAS adapted itself to the massive electoral majority and the wave of reform decrees, offering support and muting criticism. It described its main effort as one of attempting to hold the government to its promises. Pérez' shift to entrepreneurial policies forced the MAS more clearly, though reluctantly, into opposition. MAS emphasis shifted toward developing activities among the AD rank and file, promoting the contradiction between the nationalist-populist ideology of the party and the leadership's obvious embrace of big business. The principal idea of the MAS is to insert itself into the breach between AD leaders and followers—a laudable and necessary tactic.

Within the MAS's drive to establish itself as a serious Venezuelan party concerned with the day-to-day problems of the masses and in its effort to communicate effectively in a language that can be understood, one detects, however, an increasing accommodation and adaptation to populist demands rather than an effort to instrumentalize these toward a larger socialist transformation. Rather than developing immediate concrete struggles as transitional features toward creating alternative political and social structures, the MAS appears to be widening the space that it possesses within the capitalist establishment.

The efforts to gain mass support via immediate struggles and the larger efforts involved in a societal transformation are separated into distinct stages tied loosely together through the medium of socialist education. Whether this separation of the two spheres will crystallize and transform the MAS into an earlier version of AD (vintage 1945) that refurbishes and extends the welfare state or whether it is merely a particular conjunctural feature that will be overcome in the course of the struggle remains to be seen. In the short run, the growth of the MAS will depend on its capacity to overcome its image as a party of the professional-university strata—precisely its initial strong point could be its weakest link in the future. For if the above analysis is correct, the main burden of the Venezuelan development model will fall on the working class and peasantry. The growth or demise of the MAS will depend on its capacity to insert itself in the class struggle, an effort that will be strengthened to the degree to which its theoretical perspective shifts from dependency models to class analysis.

The MIR began the 1960s as the major political force on the Venezuelan left, numbering thousands.[68] Under the combined impact of harsh repression and erroneous *foquista* tactics, the MIR shrank to an insignificant force of 40 members divided into four factions by 1970. Its lowest moment, however, was also the turning point in its recovery. By 1975, it had once again become a national political force with hundreds of active cadre, offices in all the major cities, and, more important, increasing influence in the mass organizations (trade unions, student organizations, and peasant groups). Smaller in size than the MAS, the MIR makes up for numbers with an aggressive oppositional approach to the Pérez regime that increasingly is bringing it into competition with the MAS for hegemony on the left. The MIR has, like the MAS, reevaluated its previous positions and has chosen to take the path of mass struggle, counting on the radicalization of the working class and seeking influence in the trade union movement. Unlike the MAS, however, the MIR is not ashamed of its guerrilla past nor does it downplay the extraparliamentary careers and backgrounds of its combatants to gain respectability. The MIR displays less of the MAS's caution and prudence in confronting the Pérez government. This resulted in its attracting some of those individuals who became disillusioned with the entrepreneurial approach, while the MAS was still defining it as a "reformist regime" with good and bad policies.

In rural areas the MIR has developed a series of immediate demands—closely following MAS tactics—that has gained a hearing among poor peasants who have yet to see the oil money (which has gone to the cattle ranchers and large commercial farmers). In the urban areas, the MIR was less successful in its efforts to promote local democratic power among the inhabitants of the ranchos. Their demands were too radical, the communities were easily eradicated by the regime and the constituents were too transitory for successful encadrement. In the trade unions the MIR appears to be making more headway. In the iron workers' unions and in manufacturing industries, MIR militants are winning elections to office.

Nevertheless, the rapid growth and multiple successes of the MIR will face serious challenge from the repressive arm of the regime to the extent that it is capable of effectively organizing mass demands. The crucial test for the MIR will be its capacity to sustain a following when confronted with the inevitable repression and when patronage handouts occur at election time. The MIR's turn toward the working class and peasant unions is a highly significant and reasonable move in that it coincides with major developments within the society and the central cleavages that are emerging. Its critical opposition to the Pérez regime rooted in its class analysis provides it with the ideological basis for capturing the disenchanted masses. But ideological correctness, while necessary, is not sufficient. Unless the MIR inserts itself in the working class and develops roots in the trade union movement, it can be easily beheaded. Like the MAS, the MIR must move beyond its intellectual foundations or become a captive of political impotence.

CONCLUSION

The prospects are for an industrial expansion far below what was warranted and anticipated by the gains to the state through the nationalization process. Moreover, the growth of industry is accompanied by heightened inequalities and a widening of the income differences between the owners and directors in the technologically advanced new growth sectors and the workers and service employees in the economically and socially marginal parallel economy whose vegetative existence is the best indicator that it is a mere sop to the unemployed rather than an effort at providing meaningful, productive, and adequately paying employment. On the other hand, a new thrust and vitality will be evidenced in Venezuelan entrepreneurial circles. In industry as well as agriculture, new wealth will mix and compete with existing entrepreneurial elites, fashioning a heterogeneous but increasingly clearly demarcated capitalist class with a broad set of interests tied to regional as well as local and U.S.

interests. The consolidation of the Venezuelan capitalist class will be accompanied by the extension of Venezuelan capitalism downward into the Andean region, as well as upward into the Caribbean islands and Central America. In other words, the consolidation of Venezuelan capitalism will not be accompanied by internal reforms but by production for increasingly wealthy upper and upper-middle income strata and via external expansion—finding outside markets and investment outlets to compensate for the constraints imposed by an overexploited and marginalized labor force.

The major factions of the bourgeoisie include the private finance groups, the (state) bureaucratic sectors, and the agricultural-commercial elites (plantation and cattle owners). Venezuela has ceased being a dual economy with a highly differentiated ruling class based on different forms of exploitation. Today, and increasingly so in the future, in the cities as well as the country, capital dominates the relation between owners and employees—squeezing out noncommercial relations and personal bonds. Within this developing pattern the state and the private banks increasingly facilitate the growth of large-scale production; and AD increasingly becomes the party of capitalist farming, marginalizing their former clients in the agrarian reform sector who are maintained on limited doles, primarily as electoral fodder.

Within this framework of state-promoted capitalist expansion and consolidation on sectoral, national, and regional bases, the growth of social inequalities should increase the social and political distance between the mass electoral base of the AD and its party leaders. New social alignments could result, adding new entrepreneurial support to the AD party in government, but simultaneously causing its popular base to erode (provided there is a viable alternative capable of inserting itself in the week-to-week struggles over the allocation of income and social services). Given the economic resources at the disposition of the government and barring any cataclysmic shift in oil prices or demand (possible only with the breakup of OPEC), the possibility of a sharp and massive downward shift in living standards among the masses is not likely. Rather a gradual deterioration subject to occasional welfare doles, salary increases, public works, and home building prior to elections should prevent a rapid and massive growth of the left, especially in its current fragmentary state. Nevertheless, the gap between AD's populist program and its current entrepreneurial developmentalism should lead in the near future to a revival of a significant left presence that, over the next several years, may provide a real alternative to capitalist politics. The first major indication of a general realignment of class forces took place in the newly formed iron miners' union election when two left-wing groups, the MAS and the MIR, together won more votes than AD. Clearly, AD will have difficulty holding the working class base as it executes its turn toward full-scale promotion of national capitalist development.

For the left to insert itself effectively in the current Venezuelan realities, it would have to shed the past image of Venezuela as essentially a captive nation subject to imperialist powers, whose economic mechanisms are too clogged to grow. The process of growth is well under way (despite massive corruption and ineptness) and the Venezuelan state and capitalist class are in the driver's seat: the decisions to associate or not with U.S. or European enterprises will be made by these groups in terms that they negotiate. To continue to perceive the political struggle principally in terms of the agreements that the national bourgeoisie and its state negotiate with imperial capital is to overlook the fact that it is and will be Venezuelan capital that will receive the bulk of surplus value and the Venezuelan state that is channeling the huge surplus accruing from the oil wealth. For the left to continue the nationalist outlook of the 1960s is to be a prisoner of the past; it is to overlook the significance of the nationalization issue, which for the first time makes the Venezuelan ruling class the major exploiter in the country. Because Venezuelan capital will be the central exploiter, the nationalization introduces a qualitative change in the nature of the political strategy of the left. The key issue today is over who, within the national class structure, controls the surplus. And that issue can only be decided by the class struggle, which today has center stage in Venezuelan politics.

No doubt there will be new and durable ties among Venezuela's industrialists, state entrepreneurs, and foreign capital. No doubt the outward flow of surplus will continue, albeit from other, nonpetroleum sources. But if the nationalization measures and the subsequent development projects of the government mean anything (even if only a fraction are realized), it is that the capitalist class has developed its own sources of finance, markets, and production (basic industry). Henceforth, dependence becomes a relatively secondary phenomenon that may or may not exacerbate existing conflicts, but is not determinate of the level of exploitation.

The left can insert itself into that struggle at the point of production— through the trade unions, peasant struggles, and so on—or it will, by remaining as the parliamentary "bad conscience" of a bourgeoisie that never fulfills its nationalist promise, become the loyal but irrelevant opposition. For when the mass of labor is exploited predominantly by the new infusions of capital from the Venezuelan state, nationalist denunciations of sectoral alliances and agreements between Venezuelan national and imperial firms have little relevance: it neither lessens exploitation nor improves working conditions. In fact, it may appear as if the left is offering advice to the new national owners on how to negotiate more favorable terms with their imperial partners.

The class problematic, then, is the final denouement of the nationalization process that begins with a period of state-promoted national industrialization, accompanied by massive corruption, speculation, waste, and the proliferation of new bureaucratic agencies. Within this maze the central fact is that Venezuelan society will increasingly become a clearly defined capitalist class society —with a clearly recognizable working class and bourgeoisie—even if the bloated service sector continues to expand and absorb productive resources. It is within this more clearly defined class structure that left parties can work for a revolutionary transformation of Venezuelan society.

NOTES

1. Clive Jenkins, *Power at the Top* (London: MacGibbon and Kee, 1959).
2. See James Petras, "State Capitalism in the Third World," in *Development and Change,* 8, no. 1, January 1977, pp. 1–17.
3. On Mexico, see *Latin American Perspectives,* vol. 2, no. 2, 1965, special issue; on Turkey, see Dogu Ergil, "From Empire to Dependence: The Evolution of Turkish Underdevelopment" (Ph.D. diss., State University of New York at Binghamton, 19-); on Egypt, see Mahmoud Hussein, "Class Conflict in Egypt: 1945–1970," *Monthly Review,* 1973.
4. *Síntesis Orgánica de la Estrategia General del Desarrollo* (Caracas: Oficina Central de Coordinación y Planificación [CORDIPLAN], May 28, 1975), p. 33.
5. Organization of American States, "Recent Developments in the Venezuelan Economy," CEPCIES, ad hoc group on Venezuela, May 1975, p. 42.
6. *Business International,* March 14, 1975, p. 85.
7. *Business International,* February 5, 1975, p. 45.
8. *Business Latin America,* August 6, 1975, p. 254.
9. *Background Paper: Business International Venezuelan Government Roundtable,* November 10–14, 1974.
10. *Business Week,* October 13, 1975, pp. 56–57.
11. Ibid.
12. *El Nacional* (Caracas), June 25, 1975, p. D-2; see also Gastón Parra, "Venezuela: La Expropración Petrolera," *Problemas del Desarrollo,* no. 21, 1975, pp. 91–113.
13. Interview 1 with high official in CORDIPLAN.
14. Ibid.
15. Ibid.
16. Ibid.
17. Interview 13 with high official in the BCV.
18. Interview 5 with high official in the FIV.
19. Interview 12 with high official in the BCV.
20. Interview 7 with president of Fedecámaras, Antonio Diaz.
21. Ibid.
22. Ibid.
23. Ibid.
24. *El Nacional* (Caracas), July 1, 1975, p. D-8.
25. *El Nacional,* July 8, 1975, sec. 10, p. 13.
26. *El Nacional,* June 27, 1975, p. D-10.
27. *El Nacional,* June 18, 1975, p. D-2.
28. Ibid.
29. *El Nacional,* July 19, 1975, p. D-3.
30. Ibid.
31. Quoted in *El Nacional,* July 26, 1965, p. C-2.
32. Quoted in *El Nacional,* July 1975.
33. *El Nacional,* July 29, 1975, p. C-1.
34. Ibid.
35. Ibid.

36. *El Nacional,* July 10, 1975.
37. See James Petras, "New Perspectives on Imperialism and Social Classes in the Periphery," *Journal of Contemporary Asia* 5, no. 3 (1976): 291–308.
38. *World Bank Annual Report,* 1975; *Quarterly Economic Review of Venezuela,* no. 2, April 1975; *IMF Survey,* June 9, 1975, p. 176.
39. *Latin America Economic Report,* October 3, 1975, p. 155.
40. *Latin America Economic Report,* October 17, 1975, p. 164.
41. Ibid.
42. *El Nacional,* July 20, 1975, p. D-19.
43. Interview 5 with high official in FIV.
44. *El Nacional,* July 19, 1975, p. D-1.
45. *El Nacional,* July 24, 1975, p. D-15.
46. *El Universal* (Caracas), July 23, 1975, p. 1–10.
47. *El Nacional,* July 24, 1975, p. D-9.
48. *El Nacional,* July 31, 1975, p. D-1.
49. See Meir Merhave, *Hacía una Política de Desarrollo Agrícola y de Cámbio Estructural Orientada Hacía el Exterior* (Caracas, 1974): for a statement of the entrepreneurial approach to agriculture, see René Dumont's (once described as a socialist critic of the Cuban revolution) report in "Informe de René Dumont Sobre la Agricultura en Venezuela," *Resumen,* May 11, 1975, pp. 22–29. Whatever their left credentials, both writers clearly adapted to the exigencies of the Venezuelan capitalist class, designing developing strategies to suit their needs.
50. *El Universal,* December 28, 1974, p. 8.
51. *El Mundo,* March 31, 1975, p. 3.
52. Interview 13 with official of the BCV. At the end of April 1975, President Pérez announced economic measures "guaranteed to cut deep into Venezuelan workers' living standards." Prices for most agricultural products were increased, the wheat subsidy was terminated, and bank loans were made much harder to obtain. According to the country's three largest trade union federations these measures would significantly depress the real earnings of the working class during a period of increasing inflationary pressures. See "Venezuela: Unions of Convenience," *Latin America,* May 14, 1976, pp. 150–51.
53. Freddy Muñoz and Alonso Palacios, *Consideraciónes Políticas en Torno a un Año de Gobierno de Carlos Andrés Pérez* (mimeo.) (Caracas, April 1975), p. 3.
54. *El Nacional,* July 30, 1975.
55. *El Nacional,* July 29, 1975, p. D-7.
56. *El Nacional,* July 14, 1975, p. D-1.
57. *El Nacional,* July 22, 1975, p. D-1.
58. Ibid.
59. Informal interviews with executive officer in the Defense Intelligence Agency and senior officer in the Venezuelan armed forces.
60. *El Nacional,* July 22, 1975, p. D-4
61. Ibid.
62. Ibid.
63. Ibid.
64. *El Nacional,* July 21, 1975, p. C-3.
65. Interview 5 with high official in the FIV.
66. Interview with Senator Jesus Paz Galarraga of the MEP.
67. Interview with Freddy Muñoz, a national leader in the MAS, was very helpful in understanding the evolution of the organization; see also the self-definition, "Asamblea Nacional Movimiento el Socialism" (Caracas: MAS), September 26–28, 1974.
68. Interviews with Américo Martín, Hector Pérez and Moise Molero, national leaders of the MIR, were important in tracing out its formation and new orientation.

CHAPTER 3
NATIONALIZATION AND U.S. POLICY

INTRODUCTION

Since World War II, the U.S. position as a dominant imperial power in a competitive world capitalist system has been largely rooted in the continually growing opportunities for capital expansion and accumulation through the exploitation of social classes throughout the different areas of the world economy. Ex-Secretary of State Henry Kissinger succinctly summed up the issues:

> The international economic system has been built on these central elements: Open and expanding trade; Free movements of investment capital and technology; Readily available supplies of raw materials; and Institutions and practices of international cooperation.[1]

Any fundamental challenge to the arrangements that compromise this "present economic system" is regarded by U.S. policy makers as inimical to basic U.S. interests. "America's prosperity," according to Kissinger, "requires international economic stability."[2] Such stability is equated with the maintenance of the basic features of the "present economic system."

Beginning in mid-1973, Kissinger and the National Security Council (NSC) assumed the primary responsibility within the U.S. imperial state structure for defining the contours of U.S. international economic policy. The fundamental importance of economic issues in foreign policy was clearly recognized by Kissinger: "It became clear that almost every economic policy had profound foreign policy implications."[3] The decision to centralize the making of foreign economic policy with the NSC was taken at a time "when it became apparent that the increasing power of the Arab oil-producing countries could jeopardize [Kissinger's] delicate negotiations in the Middle East."[4] The oil embargo of the United States, adhered to by almost all the members of OPEC during the October 1973 Arab-Israeli War, "exposed the dangerous vulnerability" of the U.S. economy stemming from its increased dependence on external sources of petroleum:

> The oil embargo and the series of massive oil price increases which followed underscored the degree to which we had lost control over the price of a central element of our economic system. We also found that our own economic well-being and security were threatened by the energy vulnerability of our allies....[5]

According to a recent U.S. congressional study of the domestic and foreign oil supply outlook, current oil imports (approximately 37 percent of total domestic requirements) can be expected to triple by 1977.[6] Furthermore, the disruptive impact of any future OPEC oil embargo on the United States cannot be disassociated from the repercussions of such an embargo on the major Western European capitalist countries and Japan, where dependence on imported oil ranges from 95 to 100 percent (with the exception of the Netherlands—70 percent) of total internal needs:

> Serious oil embargoes would shatter Western Euorpe and Japan whose current dependence on petroleum greatly exceeds our own. Their sources are almost exclusively external.... Severe sanctions by oil producing countries thus would involve vital interests at a very early stage, "strangling" Nippon and NATO in every sense of that word. Political, military, economic, and social interests in America would suffer accordingly.[7]

It is clear that increasingly every foreign policy measure has profound economic implications. U.S. involvement in the world economy has reached the point at which a substantial part of internal development is tied to an external dynamic. Economic expansion on a world scale has become a necessity for the growth of most major U.S. corporations. Inextricably, foreign policy and the international economy have fused to become central concerns of U.S. decision makers. Foreign policy makers' concerns range from providing loans in order to open areas to new investments to providing arms and supplies to procapitalist forces intent on overthrowing nationalist regimes. The scope and depth of commitment of U.S. economic involvement on a world scale have been paralleled and promoted by the activities of the imperial state. The purpose of the foreign policy of the imperial state has been to facilitate the most favorable terms for U.S. economic expansion, not infrequently to the disadvantage of capitalist competitor nations and Third-World countries.

Third-World efforts to equalize the terms of exchange with the industrialized capitalist countries, principally through OPEC, generated a hostile response on the part of U.S. policy makers. Basing policy on the ideology and structure of the imperial status quo, U.S. spokespersons sought to maintain their privileged position in the world economy:

> This challenge [to the present world economic structure] finds its most acute and articulate expression in the program advanced in the name of the so-called Third World. This calls for a totally new economic order, founded on ideology and national self-interest.... The objective, as with the oil price increases, is a massive redistribution of the world's wealth. The United States is prepared to study these views attentively, but we are convinced that the present economic system has generally served the world well.[8]

The U.S. response to this threat to its globally dominant position was shrouded in vague efforts to find a new basis for reasserting its supremacy. Kissinger recognized these groping efforts when he observed that "America's position could not continue in a chaotic world economy."[9] The oil challenge, insofar as it was located in Latin America, was concentrated in Venezuela. In confronting this challenge and in the subsequent evolution of Washington's policy toward the economic nationalism of the Pérez government, policy makers located this major U.S. petroleum supplier within a global as well as a hemispheric and bilateral context.

Analyzing the behavior of the U.S. imperial state toward various types of economic nationalism in the Third World requires a framework that allows focus on those factors that U.S. policy makers view as central in devising policy. First and foremost, U.S. policy makers anchor nationalist policies within the context of the class structure and political process. U.S. policy makers ask: What is the impact of the new policies on the distribution of political power and class struggle? Do the nationalist measures represent an effort to constrain or restructure capitalism? Is the new regime operating within the confines of the existing class structure or attempting to change it? Specific actions, for example, nationalization of foreign-owned properties, are evaluated in terms of these larger systemic issues.

It is important to distinguish between the primary interests of particular multinational corporations and the U.S. imperial state. The multinational corporation is overwhelmingly concerned with whether or not a particular country provides it with expanding opportunities for capital expansion and profit-taking opportunities. On the other hand, the imperial state is most attentive to the nature and context of change (structural versus sectoral issues) and to such associated issues as the impact of change on trade relations and access to strategic raw materials. The imperial state becomes directly involved on the side of a multinational corporation in its conflict with a host government under conditions in which the multinational is unable or incapable of gaining an objective that coincides with the terms of the imperial state. For example, compensation raises important questions about the nature of overall change for all foreign investment in the country and throughout the world. Likewise, where a regime places in jeopardy major U.S. investments and where local class allies are too weak to offer concrete support (when basic issues regarding the health of the capitalist system are at stake), the imperial state and the multinationals pursue the same political positions. As one State Department official noted:

> If [a country] is doing something grossly in violation of its own laws, or international laws, then we are prepared to go in. . . . If there is a serious problem that a particular company can't handle then we will come in. Where we get involved is where local remedies seem to be exhausted and a company comes in and asks for our help.[10]

It has not been difficult for U.S. policy makers to identify international law with the pursuit of U.S. corporate interests, nor, for that matter, to overrule local courts regarding interpretation of laws of their own.

The issue of "swift, adequate, and effective" compensation for expropriated U.S. investors continues to be an important determinant of the U.S. response to economically nationalist regimes in the Third World. Satisfactory compensation is a critical index of the nature and content of overall change. A regime that compensates satisfactorily does so because it is prepared to pursue further economic relations with the United States. Limited nationalization well compensated plays an important role in the imperial state decision to adopt a policy of accommodation or conflict. As regimes that nationalize are fully aware ("anticipated reactions") of U.S. policy, the decision to compensate is indeed a signal that nationalization is not meant to challenge the United States but to deal with a limited area of the economy. U.S. policy makers have pointedly made their position clear to the Third World and others who harbor nationalist sentiments:

> . . . we should continue to seek full national treatment for U.S. investment abroad, and we must insist on prompt, adequate, and effective compensation in the few cases of nationalization. Where needed and appropriate, we will bring to bear available political and economic influence to get a satisfactory resolution, recognizing that the basic sanction is the damage the host country does to his future investment prospect.[11]

In relating this issue to the Pérez government in Venezuela, one U.S. official offered the following comments:

> If the squeeze is really put on some American companies they would leave. If they come into us and complain bitterly about not receiving adequate compensation, sure we would get involved. It depends on how hard the Venezuelans squeeze.[12]

As the situation later developed, it became clear that a compensation squeeze was not on the Pérez agenda:

> I don't think we have any problems at all with the direction Pérez has gone. The iron ore companies are reasonably happy. From the official standpoint, if we don't get more than book value then there are problems on the part of the U.S. government. But the pressing thing is to continue good economic and political relations with Venezuela. We have full confidence and trust that the Venezuelan government will negotiate seriously [with the oil companies] and come up with a reasonable settlement.[13]

The "confidence and trust" of U.S. imperial officials was well placed: Pérez not only paid a substantial compensation but provided new areas for foreign activity within the bounds of state enterprises.

While the imperial state was not prepared or interested in full-scale confrontation over limited losses sustained by particular firms, it certainly did not view even the losses by individual firms as totally acceptable, especially when the problem affected the oil corporations. Limited pressure to provide for the most favorable outcome was the initial response. While serious thought was given by policy makers to proposals that would avoid the picking off of individual firms, massive retaliation by the imperial state was not an appropriate response to piecemeal nationalization, but neither was tacit acceptance. The pull and tug over appropriate ways to relate larger issues of systemic importance to the particular interests of individual firms continues to be debated, with some officials adopting a position supporting closer scrutiny and more direct involvement by the state in each conflict and others proposing a more flexible position by seeking new economic ties rather than defending old ones.

The selective nationalization of foreign-owned properties by the Pérez government did, however, complicate U.S.-Venezuelan relations. From the point of view of U.S. policy makers, these actions exposed the need for "a more coherent policy [on nationalization], particularly vis-a-vis oil companies, than we have now." Prior to nationalization, various options were under consideration throughout the Executive branch:

> One way is where you really take a hardline approach to expropriation and follow through literally; when compensation is less than book value we would have to take certain actions—which a lot of people in State and Treasury would like to do. Another approach is to say that in doing this we

would create a difficult atmosphere for the U.S. companies and the U.S. government. We would say that we accept book value, we are not complaining, but say that the companies are going to have to be aware that this is the new rules of the game with the consequences that this would have for foreign investment. Some others would argue that the government should negotiate and get into the process. Companies that are involved in resources critical to the national interest, on this view, should be taken over by the U.S. government and controlled more aggressively. This is a beginning period of evaluating how we are going to approach the whole issue of the MNC's.[14]

Within the imperial state there was increasingly expressed concern "that the oil companies are no longer a match for the producer governments."[15] This issue was taken up and debated in the National Security Council in May 1975 during the council's annual country assessment and strategy paper on Venezuela. The then U.S. Ambassador to Venezuela, Robert McClintock, proposed that the U.S. government "take a leading role in direct negotiations with Venezuela concerning oil. . . ."[16] However, the NSC as a whole rejected this position in the case of Venezuela and decided, on the basis of an appreciation of the class nature of the expropriation, that a policy of nonconfrontation would be followed:

The general policy issue of whether or not we should be dealing on a government-to-government basis has surfaced on Venezuela and oil generally. Both in terms of the general question and the specific country, i.e. Venezuela, the Treasury has been involved in the discussions. Country Assistance and Strategy Papers in the National Security Council are short-term periodic reviews of our relations with each country—government-wide reviews of our policies. After the question had been discussed, just about everybody agreed that there wasn't much to be gained by going government-to-government.[17]

The political context of nationalization—the social nature of the regime pursuing policies of state ownership, the area(s) of the economy subject to nationalization, and the extent of nationalization—determines the U.S. response to Third-World governments' efforts to implement national development projects. In Venezuela, the nationalization of U.S. properties occurred as part of a nationalist-capitalist (nonsocialist) development strategy and in a way not antagonistic to basic U.S. interests located in industry, trade, banking, and even oil and iron ore. In Chile, under Allende, nationalization of U.S. properties in the context of an anticapitalist (socialist) developed strategy generated a hostile and conflictive response on the part of U.S. policy makers. The differential U.S. responses to Chile and Venezuela are rooted in the internal politicoeconomic differences and explain why in one instance the conflict was negotiable and in another it was not.

Washington has not strongly resisted Venezuela's sectoral nationalization of particular resource areas because it was accompanied by the opening up of other downstream areas for foreign capital investment. The State and Treasury departments recognize that nationalization, rather than making inroads into profits, provided an environment for capital expansion and accumulation. Once the iron ore and petroleum nationalizations appeared inevitable, Washington moved swiftly to ensure that the terms were acceptable to a coalition of bourgeois elites (U.S. and Venezuelan). The expropriated companies were offered satisfactory package settlements—compensation and future profit-taking opportunities in oil and iron-ore-related areas. U.S. businessmen recognized that foreign capital was to be regulated, not eliminated, in order to provide space and opportunities for national capital to develop and flourish. Both private and public U.S. officials have come to recognize that Pérez promotes a foreign policy that is occasionally critical but accommodating and a domestic policy of promoting national private capital expansion and the diversification of foreign investment into the nonenclave areas of the Venezuelan economy.

In this context, U.S. strategy began as one of negotiated conflict. That is, the initial response to the Pérez government, during its populist phase (April-October 1974), contained the elements of a less flexible, more hardline, position. A World Bank official has offered some insights into U.S. policy during this period:

> When Kissinger met with the Venezuelans in Mexico [in February 1974] he was hard on the Venezuelan chancellor of foreign affairs, Calvani. He was very paternalistic with all of the other chancellors, but whenever he talked to Calvani I noticed a bitterness and hardness, and the point of Kissinger was to stress the danger of high oil prices for oil producing countries, in the sense that there would be substitutes for petroleum in a not too distant period of time. At this meeting in Mexico in Feburary 1974, at which I was a member of the——delegation, the U.S. took a hardline position against Venezuela.[18]

The resultant Venezuelan decision to compensate expropriated U.S. investors, encourage foreign capital expansion into and exploitation of nonenclave sectors of the economy, and ignore the 1973 Arab oil embargo contributed to a definitive shift in U.S. policy: support for the development of new ties rather than efforts to restore old ones.

Other factors also contributed to the changing U.S. position: Venezuela's emergence as "an important player in the world of the new international economy [as a consequence of its oil wealth]";[19] increasing U.S. dependence on Venezuelan oil in a period of global scarcity (oil imports rose by almost one third between 1972 and 1973);[20] Pérez' decision to recycle oil revenues, via bilateral and multilateral agreements, in such a way as to buttress and stabilize pro-U.S. regimes and expand the possibilities for trade and investment in semiperipheral arenas of the world economy (Latin America) where the United States is hegemonic; and Venezuela's limited anti-imperialism, which is primarily confined (in terms of practice rather than rhetoric) to developing national capitalism and changing the terms of exchange. Finally, U.S. policy makers have had, and continue to have, few if any access points within the Venezuelan state and society from which pressure could conceivably be exerted on Pérez to reverse specific policies, such as nationalization. On the nationalization issue, the United States lacked any significant internal political base rooted in right-wing elements once the AD decided to take the initiative. Unlike Allende's Chile, Venezuela was not vulnerable to an external economic squeeze policy on the part of the U.S. government in collaboration with the so-called international banks. Petrodollars eliminated any possibilities, in the medium term, of external financial dependence. Therefore, on the basis of a minimal area for maneuver and the Venezuelan decision to maintain ongoing ties and create new areas for capitalist expansion, the U.S. government devised a policy of accommodation and negotiation over areas of limited conflict.

The limited areas of conflict within the relationship primarily revolve around the issues of oil prices and trade preferences. U.S. policy makers continue to criticize, explicitly and implicitly, Venezuela's leading role in pushing for oil price increases within OPEC. In July 1975, Treasury Secretary William Simon, without mentioning Venezuela by name, placed the responsibility for the decision of the major U.S. oil companies to increase oil prices on the OPEC cartel and raised the possibility "that the United States might take economic and financial countermeasures if the 13-nation cartel of oil-producing nations raised prices again."[21] The specter of retaliation was again alluded to by Kissinger in subsequent congressional testimony.[22] In September, President Gerald Ford singled out Venezuela for specific criticism over its advocacy of oil price increases in a note to President Pérez. In the same communication, Ford also voiced displeasure at Venezuela's support, at least verbally, for anti-imperialist-nationalist measures in Latin America, principally referring to the Panama Canal issue.[23] Nevertheless, the Executive branch has continued to oppose actively Venezuela's exclusion from the generalized tariff preferences of the 1974 U.S. Trade Act on the grounds that such restrictions deny the U.S. "tactical flexibility" at the policy level.[24]

The Pérez regime, for its part, has gone out of its way "to avoid confrontation" with the United States of the sort that would signal a fundamental rupture in the ongoing relationship.[25] The limits of the Venezuelan position are reflected in the differing attitudes towards tariff preferences (secondary) and oil supplies (primary). On the one hand, Pérez is prepared to irritate U.S. policy makers with an open statement to the New York *Times* declaring that Venezuela's exclusion from trade preferences "constitutes a clear act of economic aggression and political pressure."[26] On the other, Pérez has insisted that Venezuela will remain a reliable source of oil supplies for the United States and the industrialized capitalist world: "As far as Venezuela was concerned ... the production of oil must take into account the needs of industrialized countries, as much as those of any other countries...."[27] Pérez has opposed calls for a measurable reduction in the level of oil production, which, in the opinion of some knowledgeable observers, "amounted to a clear recognition that the United States, in particular, could not be pushed too far, and is in line with Henry Kissinger's recent warnings to oil producers."[28]

U.S. policy has been one of making tactical adjustments on specific issues, while attempting to maintain the "present [international] economic system." In the case of Venezuela, this has involved an acceptance of certain modifications in the price of oil. But beyond its success in regulating oil prices and partially changing the terms of exchange, the Venezuelan government has basically supported (in practice) the U.S. contention that the "present economic system [based on expanding trade, unrestricted movements of investment capital and technology, reliable supplies of strategic raw materials, and so on] has generally served the world well."[29]

The overall U.S. policy toward the bourgeois AD regime is one of accommodation and negotiated conflict. At the operational level, the U.S. imperial state agencies share this common perspective, although different executive agencies play specific roles in realizing or pursuing this policy. Within these specific functions, agencies may have overlapping responsibilities, as well as separate interests rooted in the functional institutional structure. Consequently, one finds a convergence of efforts side by side with the emergence of specific bureaucratic conflicts. But these specific differences on particular issues are located within the framework of the larger consensus over policy goals. In the case of Venezuela, these differences have not challenged the definition of what constitutes the larger U.S. interest.

The central bureaucratic conflicts center around three Executive branch agencies: the Department of State, the Department of the Treasury, and the Department of Defense. One State Department official discussed the functional institutional issue concerns of State and Treasury, out of which emerge different priorities:

> Treasury tends to be more domestically oriented, more concerned about what to do about the high petroleum prices. Treasury is more ready than the State Department to take action on the high oil prices. Treasury tends to reflect the domestic conservative financial viewpoint and gives less weight to foreign policy than the State Department.[30]

Nonetheless, the same State Department official maintained that there was no strong interagency conflict regarding U.S. policy toward negotiating and reaching agreements with Venezuela even in light of the latter's role in OPEC: "I don't think that there are any strong disagreements between the State Department and Treasury [over Venezuela]."[31] The Defense Department, again reflecting specific institutional concerns, is critical of Treasury's refusal to authorize an increase in the level of foreign military sales credits to Venezuela because of its role in OPEC. Defense officials describe OPEC as Venezuela's business and criticize Treasury for its "very short-sighted objectives—tunnel vision" vis-a-vis the Pérez regime.[32] In Defense's opinion, confrontation with Venezuela over its activities in OPEC is a short-term perspective and is less important than maintaining channels of influence with critical sectors of the state apparatus, such as the military, over the long term.

The Treasury Department accurately characterizes its particular conflicts with State and Defense as marginal ones, developing policies over specific issue areas:

> [There is] no difference of opinion vis-a-vis Treasury, the State Department and the Department of Defense regarding Venezuela. Treasury's concern about the price of oil is greater than some other departments. We are concerned about the price compared with the Defense Department's concern about supply.[33]

The consensus that exists among executive agencies regarding overall policy is primary, and the conflicts that emerge (for example, price versus supply) are secondary bureaucratic conflicts.

The central axis of U.S. policy is located in the relationships betwen the two socioeconomic systems and the role within them of private capital, especially U.S. capital. In the past, U.S. economic linkages were largely based on resource ties—the exploitation of oil and iron ore. Of late, a substantial number of new investments and trading relations have come to the fore and provide a broader framework within which to examine the tie between the two countries. The subsequent sections discuss U.S.-Venezuelan relations within a more comprehensive economic framework as a basis for evaluating the relative importance of oil and iron ore exploitation and the impact of nationalization. This discussion of socioeconomic relations forms the basis for a discussion of the behavior and orientation of the major agencies of the U.S. imperial state.

U.S. BUSINESS COMMUNITY

The key to U.S. policy making is the relationship between internal developments in Venezuela and U.S. economic and political interests. The existence of a capitalist regime that provides key raw materials for U.S. industry and investment opportunities for foreign investors is a decisive consideration in

Washington. Before the role and policies of various agencies of the imperial state can be discussed, it is crucial to consider the conditions for U.S. economic expansion under the Pérez regime.

Under Pérez, as with his immediate predecessors, Venezuela is a stable "open" capitalist country controlled by "a nationalist government with money," where direct U.S. investments (book value) in 1974 exceeded $2.6 billion, of which $1.2 billion was in nonoil activities.[34] According to a *Business Latin America* survey, the sales and profits of foreign multinationals located in Venezuela, with few exceptions, increased in 1974 in comparison with the preceding year.[35] Nevertheless, the U.S. business community's assessment of the investment climate in Venezuela during the first few months of the regime was not one of untrammeled optimism.

During the 1974 Venezuelan presidential election, the U.S. business community expressed support for the Pérez candidacy in the belief that the policies of his government would be favorably disposed toward maximizing the profit opportunities of imperial capital in Venezuela. This attitude was summed up immediately following the election: "Although he calls for a decisive state role in the economy, Pérez is not likely to continue the trend, started under his predecessor, toward increased state dominance."[36] However, the limited nationalist-populist policies of the new government between April and October 1974 served to generate a widespread sense of betrayal among businessmen. Reflecting Venezuela's emergence as a regional economic power (due to oil revenues), Pérez' initial actions were essentially directed toward redefining the terms of foreign capital penetration and expansion toward a position of greater equality with the interested multinationals:

> Foreign capital [declared Pérez] will not be welcome if it results in a monopolization of privileges, the displacement of local firms, and the use of local financial resources for non-Venezuelan firms, the payment of excess royalty or license fees, or if it dominates fundamental sectors of the Venezuelan economy.[37]

This nationalist capitalist strategy, together with an aggressive bargaining stance in the foreign economic policy area (over such issues as terms of trade and technology transfers), was, according to the U.S. business community, symptomatic of a "cavalier attitude toward foreign investment."[38] These Venezuelan measures not only meant that "no longer is the door wide open to foreign investment as it had been earlier" but also raised the possibility of Pérez following an "increasingly nationalistic" direction.[39]

U.S. corporations were especially apprehensive over Pérez' verbal support for a rigid adherence to those sections of the Andean Pact investment code providing for equity ownership of all new foreign investment within a 15-year period and the 14 percent limitation on profit remittances abroad:

> ... no matter how liberal their implementation, these regulations change the foreign investment climate in Venezuela. They mean completely new operating procedures for companies already located in Venezuela and new ground rules for companies considering the oil-rich Venezuelan market.[40]

Many industrialists had assumed that the Andean common market was primarily political rhetoric designed for domestic consumption and that, consequently, the investment code would not be rigorously enforced. U.S. businessmen also exhibited hostility toward other aspects of this progressive period: the new laws and decrees "aimed at holding down prices, lifting salaries, and encouraging production,"[41] for example, the institution of a minimum wage for industrial and agricultural workers and a decree establishing salary increases for all workers earning no more than 5,000 bolivars a month. Imperial capital responded to these feeble reform efforts by disinvesting and remitting capital out of the country, which adversely affected production goals. U.S. corporate officials also expressed some concern over the lack of skilled labor and capable middle-level administrators and technocrats as well as bureaucratic redtape.[42] The negative response of the U.S. capitalist class to the Pérez regime during this period of limited changes was a function of its hostility to specific measures taken and of its uncertainty as to the outer limits of the process of change, particularly in regard to laws and decrees affecting the profitability of foreign capital operations in Venezuela.

The period initiating a reconciliation between U.S. business and the Pérez regime, beginning in late 1974, coincided with increasing corporate recognition of the fact that, despite accelerated state intervention in the economy, the new development project was still anchored in "the free enterprise system."[43] This was most vividly illustrated in the case of the satisfactory compensation arrangement concluded between the government and the former U.S. owners of the nationalized iron ore industry in January 1975.

The U.S.-owned iron ore mines accounted for 96 percent of the total iron ore production of Venezuela, and the combined net book value of the mines was assessed by company officials at approximately $200 million, a figure considerably in excess of Venezuelan government estimates. The Pérez administration decided on compensation payments of $86.7 million to the Orinoco Mining Company (subsidiary of U.S. Steel) and $17.7 million to the Iron Mines Company (subsidiary of Bethlehem Steel), both sums to be paid over a ten-year period at 7 percent per year and exempt from Venezuelan tax. The U.S. companies agreed to accept these lower payments without significant resistance because what interested them as much as, if not more than, the level of compensation was the nature of the ongoing or future ties offered by the Venezuelan government:

In both cases a one year contract has been signed for a managerial team to remain at work in all the main installations. In addition there is a two-year "technical assistance" contract, and the purpose of both is clearly to make sure that the handover is as smooth as possible. Meanwhile both Bethlehem and U.S. Steel will continue to receive the same quantities of ore as at present —an arrangement which, according to the opposition COPEI party, "converts Venezuela into a mere subsidiary" of the two United States companies.[44]

Another important ingredient that paved the way for the "positive" nationalization of the iron ore industry was an apparent agreement "to the effect that the companies will continue to ship iron ore to their former U.S. parent companies for the next five years"[45]—a significant concession by the Venezuelans in view of the fact that the United States has developed a considerable dependence on this resource and has absorbed over 50 percent of Venezuela's total annual output during the five years preceding nationalization.

The populist measures were a transitory phenomenon of the immediate postelection period, as were the declarations restricting foreign capital. Subsequently, it became increasingly clear that sufficient incentives, that is, profit-taking opportunties, remained for foreign investors in Venezuela that more than canceled out the constraints. An official of the U.S. Chamber of Commerce put it quite succinctly:

U.S. businessmen are pragmatic. . . . What's being done with nationalization is being done with compensation. There is no question that the rules are changing, and the country has money to make the rules work. Secondly, everybody realizes that Venezuela is going to have a lot of money coming in, and the sensible thing to do is be constructive.[46]

In a situation where the U.S. government lacked the capacity to pressure Pérez from the outside; where the development strategy was characterized by sectoral, not structural, change; and where petroleum revenues had generated new opportunities for profitable activities, the U.S. business community decided to adopt a more pragmatic approach and adjust to the new rules. "Constructive" behavior superseded a conflictive posture and became the basis for the continuing relationship. As the process of clarification and nonimplementation of the regime's policies occurred, the U.S. multinationals began to move into nonextractive areas of the economy and to elaborate the basis of new ties and associations through collaboration with the local private bureaucratic bourgeoisie to share in the benefits of capital exploitation: "Companies see markets, natural resources. One corporate view is that they are not bothered by majority ownership but attracted by resources and markets."[47] The likely long-run consequences of this process, as the following tables underscore, is the reentry of U.S. capital into a dominant position in the Venezuelan economy through working from within.

Table 2 shows that the total value of U.S. exports to Venezuela increased by over $1.1 billion between 1970 and October 1975. What is even more striking, however, is that during the period 1973–75, under the impact of oil price increases and an expanded Venezuelan market, the value of U.S. exports increased by $823 million, accounting for approximately 82 percent of the total increase between 1970 and 1975. Put another way, it can be observed that the total value of U.S. exports to Venezuela between 1970 and 1975 amounted to $7.08 billion, of which $4.63 billion was accounted for during the 1973 to October 1975 period.

The rate of expansion during 1974–75 has slowed down in comparison with the staggering 70 percent increase in the value of U.S. exports to Venezuela between 1973 and 1974, but the growth pattern continues and compares more than favorably with the 1970–73 period. Among the fastest growing sectors within the expanding Venezuelan market for U.S. exports during the period under discussion, and especially since 1973, are machine industries, manufactured goods, transport equipment, and automobiles. Nevertheless, it is clear that a variety of U.S. capitalist interests have substantially benefited from the impact of Venezuela's oil price increases, as the significant increase in the value of their trade with Venezuela suggests.

The figures in Table 3 present data on U.S. direct manufacturing investment in those Latin American countries defined by U.S. policy makers as having rapid development potential. Between 1966 and 1974, Venezuela experienced a rapid rate of growth in U.S. manufacturing investment second only to Brazil. Furthermore, as Table 4 makes clear, the relative importance of U.S. manufacturing investment as compared with oil investment in Venezuela has increased significantly since 1966.

The ratio of petroleum to manufacturing investment has declined from 5.5 to 1 (1966) to 3.9 to 1 (1969) to 2.5 to 1 (1972) to approximately 1 to 1 (1974). Between 1966 and 1974, oil investment has declined from 72 to 37 percent of total U.S. investments in Venezuela, while manufacturing investment has increased from 13 to 34 percent of the total for the same period. Since 1966, U.S. investments in chemical products, food products, machinery, and transportation equipment have increased over 100 percent and in metals the increase is in the vicinity of 300 percent. In other words, U.S. capitalists have diversified their investments in Venezuela and the relative weight of new industrial-manufacturing investment is growing over time and in proportion to oil investment and total U.S. investments.

It is evident that, at this point in time, trade rather than investment is presenting U.S. capitalists with greater opportunities for maximizing profits, but it is the total nonoil stake that becomes central to any adequate explanation

TABLE 2
U.S. Foreign Trade with Venezuela, Selected Commodities (millions of dollars)

	1970	1971	1972	1973	1974	1975*
Food and live animals	79.3	82.1	109.4	127.3	243.4	163.1
Chemicals	87.8	90.6	94.3	107.7	218.5	196.4
Manufactured goods by chief material	99.9	91.9	112.0	144.5	282.8	290.8
Machinery and transport equipment	359.1	378.7	454.0	484.0	720.9	928.6
Cereals	51.2	54.7	72.6	101.9	192.1	121.4
Chemical elements and components	26.8	26.5	30.0	37.6	83.4	79.5
Iron and steel	26.7	20.1	26.4	48.6	125.8	110.1
Machinery, nonelectric	192.9	212.8	244.0	257.7	399.9	543.5
Electrical machinery	60.0	59.7	74.6	81.5	111.0	129.6
Transport equipment	106.3	106.2	135.3	144.8	210.0	255.4
Road motor vehicles and parts	84.3	88.1	109.0	127.8	176.7	210.0
Passenger cars, trucks, and so on	47.5	44.0	56.1	69.8	82.8	114.0
Motor vehicle and tractor parts	36.6	43.9	52.6	57.7	93.8	95.5
Machinery and appliances and machine parts	100.3	109.5	122.5	—	187.5	254.3
Machines for special industries	32.2	40.9	40.7	—	80.3	124.1
Wheat	36.2	37.3	49.1	67.1	124.0	80.2
Tractors, except road and industrial	8.8	7.8	13.8	13.2	24.6	57.9
Construction and mining machinery	18.1	21.1	24.6	29.2	45.9	81.0
Pumps, centrifuges, and parts	27.0	28.5	26.2	29.3	51.8	65.0
Mechanical handling machinery and equipment and parts	15.6	14.4	15.8	18.0	29.0	62.0
Total	753.7	779.6	917.5	1,023.3	1,758.0	1,846.3

*January through October of 1975.
Source: U.S. Department of Commerce, Social and Economic Statistics Division, Bureau of the Census, Washington, D.C., *U.S. Foreign Trade, Exports, World Area by Commodity Groupings*, annual 1970, 1971, 1972, 1973, 1974, 1975 (January–October).

TABLE 3

U.S. Direct Investment in Selected Latin American Countries, 1966-74, Total Manufacturing (millions of dollars)

	Countries			
Year	Argentina	Brazil	Mexico	Venezuela
1966	510	574	927	281
1967	536	627	1,016	288
1968	589	757	1,147	347
1969	659	889	1,227	378
1970	669	1,075	1,380	416
1971	712	1,225	1,492	461
1972	741	1,561	1,631	487
1973	768	2,046	1,800	517
1974	759	2,516	2,148	604

Source: Research Branch, International Investment Division, Bureau of Economic Analysis, U.S. Department of Commerce, Washington, D.C., January 12, 1976.

of the countervailing tendencies (opposing confrontation over nationalization) at work in the making of U.S. policy toward Venezuela. Here U.S. exports to Venezuela in 1974 amounted to approximately $1.8 billion and U.S. manufacturing investments during the same year totaled $604 million. Compared to the level of U.S. petroleum investments in Venezuela during 1974 ($659 million), it is probable that the magnitude and growth of the U.S. nonoil economic stake in Venezuela have been largely responsible for the evolution of a U.S. policy of accommodation and negotiation rather than hostility and conflict.*

Petrodollars have not only had the positive effect of expanding the Venezuelan market for U.S. exports but "the fact that Venezuela has petrodollars and is lending them abroad"[48] has also had important consequences for U.S. capitalists. For Venezuela has been exporting surplus capital—$700 million of an estimated $2.5 billion total commitment was disbursed in 1974—to precisely those areas of the Third World (particularly the Caribbean and Central America) where the United States is hegemonic in terms of investment capital and markets. These capital lending policies have made substantial contributions to stabilizing these dependent capitalist economies and, hence, benefited U.S. multinational corporations located in these areas.

---

*It is not asserted that one can necessarily measure political interest by some quantitative sum of economic stakes, but it does provide an indication of such an interest.

TABLE 4
U.S. Direct Investment in Venezuela, 1966–74 (millions of dollars)

Year	Industries	Petroleum	Total	Food Products	Chemical Products	Manufacturing Primary and Fabricated Metals	Machinery	Transportation Equipment	Other
1966	2,136	1,544	281	24	66	12	33	46	100
1967	2,081	1,481	288	24	68	14	28	(D)	(D)
1968	2,158	1,480	347	26	(D)	15	33	(D)	143
1969	2,196	1,474	378	31	(D)	18	39	(D)	154
1970	2,241	1,440	416	36	(D)	25	43	(D)	170
1971	2,199	1,327	461	39	101	21	(D)	69	(D)
1972	2,172	1,225	487	44	110	22	60	81	169
1973	2,051	(D)	517	45	130	32	57	74	181
1974	1,772	659	604	54	145	40	67	90	208

Note: (D) = Suppressed to avoid disclosure of data of individual reporters.
Source: Research Branch, International Investment Division, Bureau of Economic Analysis, U.S. Department of Commerce, Washington, D.C., January 12, 1976.

In summary, it is clear that, while specific areas of the economy have been reserved for national companies, Venezuela remains a relatively open society from the vantage point of the U.S. capitalist—as an expanding market for exports (machinery, transport equipment, and so on), management know-how and technical expertise, and as a location for large-scale investments in downstream ventures, such as petrochemicals, chemicals, steel, and aluminum. Venezuela is also in "the proximity of Colombia with its sizeable, low-cost labor pool that tends to spill over the common border."[49] These factors, in concert with the country's sophisticated financial network, extremely low corporate taxes, oversupply of capital, and lack of controls on local or foreign financing contribute to an ease of operations for U.S. multinationals unequaled in any other Latin American country. While locally owned firms have some advantage as regards borrowing costs and access to local credit, the multinationals view such problems as minor irritants with very limited consequences.

Venezuela's industrial expansion program for 1974 significantly expanded the internal market for U.S. capital and nondurable consumer goods, and the "temporary setbacks" resulting from the initial price controls and wage increases were eventually surmounted, due largely to the implementation of a "much more flexible" policy on prices by the Pérez government.[50] In practice, prices were allowed to increase relative to wages, with the resultant increase in profits for the local and imperial bourgeoisie. The investment climate in Venezuela at the end of 1974 was described in highly positive terms by Business International Corporation following a roundtable with members of the Pérez government in Caracas:

> ... foreign investment has been encouraged by the country's attractive investment climate. Except in a few areas such as banking and insurance, there has been no discrimination against foreign capital. Foreign-owned firms have been free to remit profits, interests, and royalty payments, and exchange controls have not been a problem.[51]

In October 1975, President Pérez emphasized the reversal of his earlier populist pretensions: "We have learned to be pragmatic. . . . There is no danger that the behavior of Venezuela will drive foreign investment away. . . ."[52] This pragmatism has manifested itself, above all, in the nonimplementation of the Andean Pact foreign investment code to the extent that "the new rules that foreign investors will be forced to live by are not much more restrictive than foreign investment laws in other parts of Latin America. . . ."[53]

The U.S. business community has opposed the exclusion of Venezuela from the generalized tariff preferences (GSP) of the 1974 U.S. Trade Act because of its OPEC membership. The Council of the Americas, an association representing approximately 200 U.S. corporations with Latin American subsidiaries (which account for around 90 percent of all U.S. private investment in the hemisphere), has actively lobbied within the U.S. government for the extension of trade preferences to Venezuela. A council memorandum summarized the views of over 180 corporate executives throughout the United States regarding Section 502 (b)(2) of the Trade Act pertaining to generalized tariff preferences:

> Most COA members support the view that the United States should take economic counter-measures against those countries that withhold supplies of vital materials from the U.S. However, they stated that since Venezuela and Ecuador have not withheld petroleum supplies from the U.S., they should not be excluded from GSP eligibility solely on the basis of their membership in OPEC.[54]

In testimony before the House Subcommittee on Trade in May 1975, Henry Geyelin, president of the Council of the Americas, pointed to the larger significance of a nonconfrontation strategy with Latin America over the issue of GSP for Venezuela at the present time:

> ... our favorable balance of payments with Latin America has been in the nature of $1.5 billion, a number which we can't say of many other places in the world. Anything which we can do to assist the developing countries in going from, if you will, a one-crop economy to diversification and industrialization is only going to mean increased sales and trade with this country as their economies expand and diversify.[55]

The American Chamber of Commerce of Venezuela presented a similar rationale, but specifically in terms of Venezuela, before the Senate Foreign Relations Committee. It objected to Venezuela's exclusion from the trade preferences on two grounds: first, because Venezuela had refused to participate in the Arab oil embargo in 1973 and had been a consistent supplier of oil throughout every major regional-global conflict involving the United States since World War II, and second, and more important, this punitive measure against Venezuela might threaten the exceptional opportunities for expanded capital accumulation in Venezuela, that is, new industrial investments and the movement of U.S. oil companies into new profit areas of the economy: "There is no question but that the industrial expansion being planned in Venezuela will generate a sharp increase in imports of U.S. capital goods, while consumer goods imports have already increased abruptly and should continue to rise."[56]

In June 1975, the U.S. Department of State offered a glowing assessment of the Venezuelan market's profit potential:

> Sales prospects for U.S. businessmen have probably never been better in the Venezuelan Market.

Venezuela's broad and ambitious national development program, the nation's increased revenue which continues to accrue from the four-fold increase in petroleum prices, and substantial credit offers by the government to the private sector form the basis of a strong market for a broad spectrum of U.S. goods and services.

The Venezuelan national development scheme includes plans to create large new enterprises engaged in a host of high technology industries, e.g., enlarged steel making and milling facilities, shipbuilding and petrochemical plants. Much of the technology required will come from abroad and it is important that U.S. engineering and design firms be especially alert to opportunities for participation.

In summary, business in Venezuela is good, trade opportunities abound across the board, substantial domestic credit is available and payments are prompt.[57]

U.S. companies were indeed "alert to opportunities for participation" in the Venezuelan development program. According to *Business Latin America's* 1975 survey of 15 foreign multinationals operating in, or exporting to, Venezuela, the trends evidenced during 1974 were maintained and surpassed: "Most of the 15 interviewed experienced strong or record-breaking gains in sales and profits last year. The range of sales increases was from 5 percent to 38 percent. Profit gains ranged from 15 percent to 68 percent."[58]

It is clear that a diverse number of major U.S. companies are developing an important and lucrative stake in the nonoil sectors of the Venezuelan economy. This stake has grown substantially, precisely as the Venezuelan state has increased its share and size of oil revenues. Furthermore, U.S. corporate executives have recognized the growing opportunities in the current development strategy and have articulated a policy that clearly accents the positive, thus influencing U.S. policy toward accommodation. The sharp increase in trade, and specifically Venezuela's increasing demand for machinery and technology, have further spurred U.S. corporate profits at a time of internal economic crisis. The basic thrust from the U.S. corporate business community was not to limit U.S. relations in order to contest nationalization but to push the U.S. to widen its relationships with Venezuela. Paradoxically, then, nationalization has led to an increasing role for U.S. corporations throughout the Venezuelan economy and increased profits for U.S. corporate interests.

OIL COMPANIES

As the major force among the U.S. business community in Venezuela, the executives of the multinational oil companies assess the relationship between the United States and the Pérez government in terms of the potential for profit maximization. Future economic relations depend in large part on essentially politicoeconomic considerations: the capacity of the regime to limit socioeconomic changes, to control political forces advocating policies of social transformation, and to promote continued profit-taking opportunities for foreign capitalists. As an executive from Exxon noted:

> The considerations for investment are not in any way unique there. Clearly, U.S. companies are looking at such basic questions as the stability of the government, the extent to which risk is involved in these institutions. These are the major considerations.[59]

One oil company executive addressed this issue with specific reference to Venezuela:

> I would have said that it is among the more stable South American countries. I would see it as quite a favorable place to invest in. Over the last few years, the government gives me the impression of having a broader basis of support. It is less likely to be turned up by a revolution. That really pinpoints it. It has a broader basis of support in the country and this must lead to greater stability compared to Venezuela itself in the 1940s and 1950s.[60]

The maintenance of a stable political order is synonymous with an ongoing investment climate that is characterized by incentives to foreign capital (tax concessions and so on), that permits the relatively unrestricted remittance of profits, and, generally, that provides a satisfactory environment for carrying out the business of business:

> Basically, in general, one looks for stability so that if one government succeeds another, the investment climate in the sense of incentives, permission to remit profits, conditions for employment and payment of staff and so on are not radically changed from one government to the next.... [In looking at Venezuela] the main concern is that in order to conduct our business we must be certain, in advance, that we can get the supplies we have contracted to deliver, and ... obviously there must be a certain amount of anxiety on our part that we shall be able to continue to secure products from them at a price which enables us to pay our way on the prices we have contracted to deliver at.[61]

There is a realization on the part of the foreign oil company executives that, given the Pérez development strategy based on creating expanded new internal markets, the possibilities for maximizing profits are now greater in the industrial sector of the economy, and that it is the nonenclave capitalists who will be the major beneficiaries of this strategy: "I'm sure that through the eyes of many businessmen the outlook in Venezuela is much brighter today than five or ten years ago. Their wealth alone must be creating new markets which are of interest to U.S. business."⁶² As regards the overall possibilities for imperial capital expansion in Venezuela as compared with other Latin American countries, an official of the Exxon Corporation offered the following comment: "There is no question that there are four places in Latin America now where there is a rather big future—Mexico, Venezuela, Argentina, and Brazil. Among these four, Mexico, Brazil, and Venezuela stand out."⁶³

Toward Nationalization

The foreign oil companies' response to the Pérez government's decision to nationalize the petroleum industry and to the legislation introduced into the Venezuelan parliament for that purpose was reflected in two seemingly contradictory attitudes. On the one hand, there was a general hostility toward the idea of nationalization itself coupled with an uncertainty as to whether any (let alone satisfactory) compensation payments would be forthcoming. On the other hand, the oil companies were aware of "the worldwide trend toward national ownership of natural resources, and national control over their extraction."⁶⁴ Thus, resistance to a government takeover of their properties existed side by side with a predisposition to "accept the fact that nationalization is the wave of the present" and adjust or accommodate to this global trend.⁶⁵ Differences emerged among oil companies, with the U.S. oil companies (especially Exxon Corporation) adopting a considerably more rigid position on the nationalization issue in comparison with other foreign oil interests operating in Venezuela. An executive of Asiatic Petroleum, a subsidiary of the Royal Dutch Shell Company, alluded to this hardline response on the part of the U.S. oil multinationals:

> We must accept the authority of a national government, and in that authority is included the right to nationalize and take possession of its national resource, provided that nationalization or reversion doesn't amount to expropriation without due compensation. Some of the American companies might be inclined to take a harder line and I think this applies not only in the present context of oil, but in general, where companies operating abroad come up against a newly established national government.⁶⁶

The hardline posture taken by specific U.S. oil companies must, however, be understood within the general position adopted by the foreign-controlled oil industry that tied accommodation to the "very basic questions of prompt, adequate and effective compensation."⁶⁷

The position of the oil companies on nationalization softened in direct relation to the terms of settlement and, more important, the continuing opportunities for profitable activities in related areas of oil production other than direct exploitation. As it became apparent that a substantial settlement would be made meeting most of the oil companies' claims, and as it became clear that some oil companies would continue to play a role in marketing and shipping and in the development of new technology, the voices of moderation and reasonableness began to dominate within the oil industry.

Beyond Nationalization

In the postnationalization period, the foreign oil companies have displayed considerable (though varying) interest in playing a prominent role in collaboration with the state-owned industry, especially in commercialization, distribution, and technology: "What interests them much more [than ownership] are processing and international marketing, where the big profits are to be made, and providing their technical experience and services."[68] In other words, the response to the fait accompli of nationalization was primarily defined by the possibilities for elaborating new relationships in areas related to oil that promise large-scale profits. Acquiescence grows out of earlier unsuccessful efforts to reverse the nationalization process due to lack of internal support and the incapacity of the U.S. government to exert leverage on Venezuela in behalf of the oil companies. For the multinational corporation, nationalization is not necessarily a decisive loss as long as it is not displaced from other areas of the economy where it can invest and maximize profits.

The foreign oil companies were generally of the opinion that the Pérez government would continue to require their assistance, in the form of joint ventures, because of the paucity of experience and know-how in the areas of marketing and technology:

> Some sort of role for foreign oil companies will have to be provided for two basic reasons: (a) Venezuela doesn't have an international marketing structure of its own. The oil bodies in Venezuela do not have a network of international marketing contacts sufficient to market 2½ million barrels of crude oil a day, i.e., short-term needs; (b) for technological input. When their reserves begin to decline seriously, when they start thinking in terms of developing new reserves, they do not have the technology they need singlehandedly without outside assistance to undertake this. Yet they have to offer attractive terms to foreign companies. As a result, some legal technique for foreign ventures seems to be required.[69]

> It will be in the best interests of Venezuela and ourselves if some sort of relationship continued so they can have access to our know-how and plant and infrastructure. We want to develop a mutually beneficial and acceptable process for continuing the flow.[70]

Similarly, *Business Latin America* argued the likelihood of new areas of profit-taking opening up to the oil companies as opposed to their total exclusion from the economy:

> Pérez is aware that despite the country's wealth and abundant natural resources, it lacks the technology and skilled manpower required to unleash its potential. Therefore, the door must be kept open to foreign technology, especially during the transition period.[71]

This fact, observed the British-based *Quarterly Economic Review of Venezuela* in April 1975, "is probably one reason, why, at present, they [oil companies] seem to be treading softly as regards compensation for their assets. . . ."[72]

While the oil companies remained hopeful of reaching postnationalization agreements with the Pérez government in the marketing and technology areas, they expressed differing degrees of optimism as to whether, in fact, the government would defer to them in these particular areas. Mobil Oil even raised doubts over the future profitability of operating in Venezuela, notwithstanding joint venture agreements. In general, the differences among the oil companies reflected the degree to which each multinational had secured concession arrangements prior to nationalization and the stake of each individual company in Venezuela vis-a-vis its investments elsewhere. An executive of Asiatic Petroleum put it this way:

> I think there are differences between the companies operating there. I think that this varies to some extent according to the concession arrangements they already have, which vary to some extent. It also differs according to the stake a company feels it has in Venezuela, i.e., how important is Venezuela as a source of crude and how much it is going to lose if it doesn't come to an amicable agreement on reversion.[73]

For those companies with concession agreements and investments on a world scale, the attitude was quite optimistic. For those companies with concentrated investments in Venezuela and/or few concessions, there was a doom and gloom attitude. Mobil Oil gave Venezuela a low priority as far as future investments and profit-taking possibilities in oil-related activities in the postnationalization period were concerned:

> Right now, Venezuela is way down at the bottom of the list. Because of the uncertainty about future contractual arrangements, we are looking elsewhere where we can count on stable terms, etc., and we would not look at Venezuela until the situation clarifies. Once terms are clarified, and once we see we can have a relatively stable, predictable future on reasonable terms compared to our other investment opportunities, then we would be prepared

to make further investments. Other companies are willing to gamble, such as British Petroleum. You always have to look at this from the point of view of comparable investment opportunities. There are other opportunities around at this time that are safer and better.[74]

In the world of the multinationals, particular investments are always located within a larger global perspective:

> It's always a global perspective that determines the final decision on any one investment. For example, it is clear that in Mobil Oil in the last six months a decision has been made at the top to stress domestic exploration in the U.S. This is a reaction to a number of pressures—the difficulties in finding satisfactory investment climates, etc. The individual decisions are determined in a competitive perspective always. The main thing is the global viewpoint and the top management's overall assessment of where at any one point our limited resources should be invested.[75]

Future joint ventures with the Pérez government are assessed in terms of comparable investment opportunities in other areas of the globe. On that criterion, according to Mobil Oil executives, Venezuela compares unfavorably with such stable, authoritarian-dictatorial capitalist regimes as Indonesia and Nigeria: "We are absolutely delighted with Indonesia and we are doing very well as far as the finds in Nigeria are concerned."[76] Finally, although Mobil Oil expressed a substantial interest in retaining its role as an "international marketer" of Venezuelan petroleum, company executives were perturbed over the "growing trend towards state-to-state sales" and the possibility that the Pérez government might follow this trend and assume the responsibility for marketing and distributing the country's oil exports.[77]

Asiatic Petroleum, the Royal Dutch Shell subsidiary involved in the marketing of derivative products from Venezuelan oil (and, to a lesser extent, the U.S.-owned Exxon Corporation), maintained a high degree of confidence regarding a profitable ongoing relationship with the Pérez government. Company officials argued that the Venezuelans accepted the fact that the existing outlets and established distribution channels provided the most efficient avenues for exporting petroleum (crude and products) and, therefore, would act accordingly:

> ... although governments can market their own crude and products, the experience of other oil-producing countries, in the Middle East, for example, seems to be that it is not quite as easy as they thought a year ago—especially in the face of a rather slack market in 1974–75—and they may continue to make use of transnational oil companies' marketing expertise.
>
> Existing outlets for crude are quite an important consideration. At first sight it is thought that crude is crude, and that you load it into a tanker and

move it and somehow all is well. This is true at the margin, but for the bulk of the oil it only moves efficiently through established channels. I believe that the Venezuelan government realizes this. Venezuela is producing about 2½ million barrels a day and they need the revenues from that in order to carry out their development plans. They still will wish to dispose of approximately that amount of crude even after reversion, and at the moment we see no reason why we should not come to a reasonable arrangement to help them dispose of it.[78]

However, these same officials were prepared to admit that Venezuela could conceivably dispose of most of its oil exports by drawing on the marketing advice and expertise, not of the multinational oil corporations, but of the large number of independent oil consultants in Venezuela:

... There are quite a number of extremely well-qualified consultancies who presently earn their living by hiring out to the oil companies and there is no reason why the Venezuelan government should not employ them direct rather than draw on the expertise of the oil companies. That doesn't eliminate the disposing of the product but even that could be done. And, therefore, the present oil compaines, it seems to Shell, are competing with alternatives open to the Venezuelan government ... on the marketing issue.[79]

The foreign-owned oil companies in Venezuela estimated the combined value of their assets at $5 billion, but agreed (with the exception of El Paso) to accept as satisfactory an offer of approximately $1 billion in compensation payments by the Pérez government.[80] One other U.S. oil company, Occidental, had its offer set aside due to charges that it had used over $3 million to bribe Venezuelan government officials and political candidates in 1970 and 1971 in an effort to gain lucrative service contracts to exploit oil deposits in southern Lake Maracaibo.[81]

Under the terms of the initial Venezuelan offer, the oil companies believed that 44.5 percent of the payments would likely be in the form of cash and the remainder in five-year bonds at a 6 percent tax-free interest. Subsequently, the Pérez government announced that only 9 percent of the total amount allocated for compensation payments would be in the form of cash and the rest in five-year bonds. "But, and this is the crucial point, the bonds themselves are redeemable only in oil, not in cash, as had been expected."[82] As a result, over 90 percent of the compensation payments will be in the form of Venezuelan oil "and since the price of oil is likely to climb higher the deal looks good for Venezuela. At the same time Venezuela will be assured of a secure market for over $900 million worth of its oil five years from now."[83]

The compensation payments were part of a package settlement that also provided for continued technology inputs into the nationalized industry as well as marketing agreements for the international transshipment of the country's petroleum exports. According to the Venezuelan minister for Mines and Hydrocarbons, Valentin Hernández, the oil companies would be offered between 8 and 15 cents per barrel for technical aid in discovering and extracting oil and a similar fee for refining technology. Transport costs were also expected to add another 7 to 15 cents to the cost to Venezuela for each new barrel of oil.[84] Some idea of the future large-scale profit-taking activities available to the foreign oil multinationals is indicated in the case of Exxon Corporation:

> Exxon's subsidiary Creole, for example, the largest oil firm with half the country's production, can nearly match its 1974 net profit of $60 million on the extraction technology fee alone, if it gets 15 cents, as expected, for each of the million barrels a day it has been producing. Even discounting salary and other costs, the addition of the other fees will mean *Exxon will make more money on Venezuelan oil than before nationalization, according to critics* (emphasis added).[85]

Whether or not the case of Exxon is typical of other oil companies the fact remains that substantial profits from oil approximating those prior to nationalization continue without the risks, labor, or political problems of the past. The oil companies' decision to accept the Venezuelan compensation offer was not only conditioned by the accompanying offer of multiple short- and long-term payoffs—principally new areas of exploitation in oil-related activities—but also by their inability, individually or collectively or in collaboration with the U.S. government, to bring sufficient pressure to bear to force an upward revision of the total compensation payments and a more favorable arrangement regarding the form of payments.

U.S. Oil Corporations and the State

The U.S. multinational oil corporations have operated, and continue to operate, in close formal and informal contact with the Executive branch of the U.S. government. "Our view of the U.S. government and the U.S. embassy," declared a Mobil Oil official, "is that they should service American businesses who are trying to invest in foreign countries, make information available, and then help them to make proper contacts, and facilitate the prospects for investing overseas."[86]

Efforts by successive Venezuelan governments since the late 1950s to gain preferential treatment for Venezuelan petroleum in the U.S. market through the medium of government-to-government agreements (which would increase the country's decision-making powers within the industry at the expense of the multinationals) were vigorously opposed by U.S. policy makers under pressure from the oil companies. For their part, the oil companies responded to Venezuelan government attempts at increasing control through such measures as increased taxes and limitations on new oil concessions by reducing the levels of investment (by almost 200 percent between 1957 and 1962), exploration (from 279 exploratory wells in 1961 to 75 in 1967), and drilling and employment in the industry (jobs declined by 28 percent between 1960 and 1966). They also expanded ties with local private capital through the aegis of the Fedecámaras (to which the petroleum industry subsequently became the largest single financial contributor) in order to exert pressure on the government through local intermediaries.[87]

Venezuelan efforts to curb the profits of the oil companies provided a rationale for U.S. policy makers' actions in support of the multinationals. As late as January 1971, for example, "U.S. officials informed Venezuela that any hopes for special preferences were 'fruitless' in view of the country's recent tax increases."[88] In September 1972, in the wake of a developing U.S. concern to formalize long-term access to Venezuelan oil, President Caldera met with U.S. officials and, once again, expressed his country's interest in negotiating a government-to-government agreement:

> The American position, however, doomed this possibility from the outset. U.S. representatives acted as if they were surrogate company spokesmen. . . . In brief, the United States asked for long-term guarantees of security, amortization, and profits for the private companies in return for company investment capital and a possible guaranteed market access to the United States.[89]

U.S. oil interests were able to influence the U.S. government to act in its interests and against the Venezuelan government in most instances. In recent years, as other U.S. corporate interests became heavily involved with Venezuela, U.S. policy makers could not pursue a hardline policy in defense of oil corporations if the results of such policies adversely affected the growing economic interests of substantial nonoil corporate groups.

Throughout the period leading up to and during the passage of the legislation nationalizing the Venezuelan subsidiaries of the foreign oil companies, the officials of the affected U.S. companies maintained constant channels of communication with, and were able to gain direct access to, influential members of the U.S. Executive branch. This interaction, as part of an effort to enlist the active support of the U.S. government against the nationalization policies of the Pérez regime, was centered within one of the key imperial state agencies, namely, the State Department. One official of the Exxon Corporation described a series of meetings and exchanges over time between company executives and State Department officials pertaining to the Venezuelan situation, including a personal briefing with the newly appointed U.S. ambassador, Harry Shlaudeman, on the problems of business in Venezuela. "It's that sort of relationship," he continued, and then proceeded to a discussion of the oil nationalization law:

> They promulgated a law a year ago that said that by the end of 1974, the government was going to take over 51 percent of our assets. They said that the government would take it over if, in the period remaining in that year, we failed to sell, [or] enter into an agreement with local enterprise to give over 51 percent. We tried, but by the deadline had not succeeded, and were taken over. At the eleventh hour we didn't ask, but implied that we wanted [Venezuelan government] intercession. We made a number of abortive efforts to contact the Venezuelan government, but they didn't respond. So we went to the State Department and laid out our story and requested that the U.S. government intercede on our behalf and get things off dead center.[90]

The U.S. oil companies' efforts in this direction were ultimately ineffective for two primary reasons: the lack of political support within Venezuela for the plight of the foreign oil companies and the incapacity of the U.S. government to exert political or economic pressure on Pérez due to the existence of OPEC and the level of world demand for this scarce strategic resource.

The oil companies were and are an important faction of the U.S. ruling class, but they do not decide policy for the whole class. In the case of Venezuela, where a partial nationalization provided for compensation and continuing profitable relations and where profitable ties were maintained in other sectors of the economy (industry, banking, exporters), the adversely affected hardline oil compaines were unable to provoke a more militant stand from the U.S. government. Furthermore, the ties and relationships between imperial state interests and the military and among the political stratum prevented precipitous action founded on the specific problems of particular enterprises. Oil nationalization appears, then, as an area of limited conflict in a double sense: limited in the number of areas or actors adversely affected and limited in the amount of resources that the imperial state was willing to commit in fighting the issue. The cost of an all-out open confrontation initiated by the United States would likely have been extremely high and, given the lack of internal and external leverage, the outcome would likely have been a disastrous failure.

U.S. STATE DEPARTMENT

The U.S. decision to negotiate rather than engage in confrontation politics with the Pérez government in Venezuela is based on the crucial distinction between sectoral nationalization that redefines areas and terms of exploitation and nationalization that is inserted into an anticapitalist development strategy. An overall agreement on the primacy of capitalist development and the maintenance of the capitalist mode of production—as distinct from conflicts over particular issue areas, such as the terms of trade or limits on specific areas of exploitation—has served to mediate a potentially antagonistic relationship. As one State Department official observed: "I don't think we have any problems at all with the direction Pérez has gone."[91] The critical determinant in the State Department's response to the Pérez development strategy is the larger politicoeconomic context within which decisions have been implemented. This factor accounts, in large part, for the different U.S. responses to the policies of Pérez in Venezuela and Allende in Chile:

> I don't think that the Venezuelan policies are at all comparable with the Chilean ones. The Venezuelans are much more interested in working out satisfactory relationships with foreign investors and specifically with the iron ore and oil companies, and even in the implementation of Decision 24 of the Andean Pact I don't think it is clear that they are trying to shackle or eliminate foreign investment. But the Allende regime was antiforeign investment. They were anticapital.[92]

In the case of Venezuela, the Pérez government is attempting to carry through a form of capitalist modernization from above in which the state plays a central role and on the basis of limited mobilization from below. In the case of Allende's Chile, not only was nationalization elaborated in the context of an effort at socialist transition but it took place within a quite different political and social context, specifically, where there existed an organized, class-conscious working class mobilized to transform Chilean society. Despite the continuing U.S. dependence on a strategic raw material resource from Venezuela, there is minimal official concern over the possibility of a future decline in the supply of oil because of the bourgeois regime that is carrying out the changes, a bourgeoisie that promotes capitalism and seeks to associate with foreign capital. In commenting on the problem of U.S. dependence on Venezuelan oil and Chilean copper, a State Department official recognized the importance of the sociopolitical context of change and the type of projected response:

> ... We have no concern over nationalization [in Venezuela] up to now. In Chile there was a lot of concern and we weren't terribly dependent upon Chilean copper. But in the case of Venezuela, there was nothing to cushion. We weren't planning to nail Venezuela to the wall.[93]

While copper was less crucial to the U.S. economy, the State Department crucified Allende for trying to implement democratic socialism; while oil is essential to the United States, the State Department is willing to accommodate a capitalist regime. What is crucial is not the extent of dependence on a resource or its importance but the socioeconomic context in which the relationship is developing.

The State Department, reflecting its primary concern with long-term, large-scale issues, assesses Venezuela and the Pérez government within a variety of contexts: bilateral (United States), subregional (Andean Pact), regional (Latin America), and global (OPEC membership and the role of OPEC within the world economy), with the latter given priority status:

> Increasingly, we are trying to consult with them as an important player on the global scene. Venezuela is an important member of OPEC, participated in the Paris [Preparatory] Conference, and is a key figure in the International Monetary Fund and the World Bank.[94]

Increasingly, U.S. policy makers became cognizant of the fact that Venezuela was both "a developing nation" and " a very wealthy one."[95] Venezuela's oil wealth accelerated the country's hemispheric or regional reach, increased its importance within the capitalist world, and, in part, allowed the Pérez government to redefine the terms of Venezuela's external dependence in the process of an emerging nationalist capitalist development strategy. The catapulting of Venezuela into an important regional, and even global, role and the decision of Pérez to commit massive financial resources, either directly (bilateral) or indirectly (multilateral), into areas of large-scale U.S. capital investment, which were also important markets for U.S. goods and services, are crucial to an understanding of the evolving State Department position on Venezuela:

> There have been changes in U.S.-Venezuelan relations on the basis of Venezuela's new-found oil wealth. From the Venezuelan side, they are feeling their oil oats. A much more influential nation, with weight in the international financial institutions. They have become a significant donor nation. The United States, late last year, began to realize Venezuela's increased importance.[96]

The initial hostile response to the nationalization policies became secondary to efforts to elaborate a strategy for channeling Venezuela's wealth in ways beneficial to U.S. capitalists and the U.S. position as a dominant global imperial power. During 1974 and 1975, Venezuela's recycling policies made significant contributions both to stabilizing dependent capitalist economies within the U.S. sphere of influence and increasing the profit-taking opportunities for U.S. investors in and exporters to these countries:

> Venezuela is very much looked at in a global context which is one of the reasons we have taken a big interest in Venezuela: (a) because of its increasing influence in the world because of OPEC, contributions to international financial institutions, and its growing leadership in the Third World and Latin America; and (b) we look upon it as a bridge to the developing nations in terms of developing a new international economic order.[97]

Future relations are determined, in large part, by economic considerations: Venezuela's role as a financial bridge to U.S.-dominated areas of the Third World and the Pérez government's support for joint ventures and the participation of foreign capital with national state and private capital in the process of capitalist development. On the basis of this analysis, the notion exists of a convergence of interests between U.S. imperial capital and the Venezuelan bourgeoisie. For the former, there is the expansion of market opportunities (export of capital goods, technology, and so on) and new areas of investment. For the latter, there is the emergence of conditions to industrialize and create its own internal, and even regional, market.

The combined impact of the Pérez government's policies together with the restricted U.S. options for reversing that aspect viewed as hostile to U.S. interests (nationalization) produced an overall State Department assessment in support of an accommodative approach. Venezuela's capital export policy received a favorable response. The fragmented and sectoral nationalization policy for iron ore and oil has neither displaced other key foreign capitalist interests nor encompassed measures threatening to the continued dominance of the capitalist mode of production in Venezuela.

On the U.S. side, there developed a recognition among State Department officials of the relatively limited capacity of the United States to exert economic, political, or military leverage on Pérez in order to reverse the new relationship. Evolving limited conflicts over such issues as Venezuela's role within OPEC and its increasing tendency to act as a spokesperson for the Third World on major social and political issues became negotiable within this larger context. According to one State Department official, these frictions must be viewed as "part of a larger dynamic relationship" in which "the United States is constantly adjusting as the situation changes."[98] This "larger dynamic relationship" is defined by the following factors: expanded U.S.-Venezuelan trade relations, new profitable investment areas within the Venezuelan economy for U.S. capitalists, and the positive consequences of Venezuela's capital export policy, especially its contribution toward bridging the distance between the rising new bourgeoisie in the Third World and U.S. imperialism.

The procedure and terms of the iron ore nationalization in January 1975 were a critical factor in the State Department's evolving overall assessment of the Pérez regime and the point of reference for the subsequent oil nationalization:

> Pérez announced the nationalization of iron ore in April 1974. We were immediately contacted by Bethlehem and U.S. Steel. Then began a period of protracted negotiations between the companies and Venezuela. On January 1, 1975, the nationalization took effect and the companies were given certain things that from their standpoint added up to a relatively satisfactory deal. Once nationalization did occur the question was how good a package settlement would be worked out. The United States did not interfere in this.

> The Venezuelan government knew our stated position on prompt, adequate and effective compensation, and they were reminded of it. They negotiated a settlement with the companies ... we were encouraged by the iron ore precedent in the case of the iron ore nationalization there is a tendency to speak more softly, and to think things would work themselves out better than worse.[99]

The "anticipated [U.S.] response" to noncompensation reinforced by "reminders" and the recent experience of Allende's Chile, together with the desire of the Venezuelan bourgeoisie to maintain good relations with the United States, resulted in a "mutually advantageous" settlement. One State Department official emphasized the centrality of compensation and "future ties" in any U.S. assessment of the nationalization equation:

> [Our] lack of concern over whether the companies will be paid is due to two reasons: (a) Venezuela has said it will pay compensation; and (b) the model is the nationalization of the iron ore industry. The two iron ore companies, Bethlehem Steel and U.S. Steel, received $115 to $116 million in government bonds. And, of course, compensation is only one part of the package. In the case of the petroleum and iron ore companies there is an interest in future relationships and supplies over the short-term. Compensation is an important factor, but so is the ongoing relationship. It all adds up to the bottom line: how much profit they make.[100]

The close relationship between the State Department and the U.S. multinationals allowed Bethlehem Steel and U.S. Steel direct and immediate access to department officials following the nationalization decree. This close relationship accounted for the efforts by State Department officials to secure satisfactory compensation and new areas of profitable activity for the iron ore companies ("the ongoing relationship"; "how much profit they make"). This reinforces the contention that economic questions are of critical importance in the elaboration and evolution of U.S. policy.

The State Department played a role in both shaping ("reminding") and influencing ("package settlement") the type of nationalization that has taken place in Venezuela under the Pérez regime. As one State Department official expressed it: "One shouldn't generalize from oil, which is a special case, to all other kinds of foreign businesses."[101]

Nationalization has enhanced the possibilities for new class ties and mutual class accommodation due to the incorporation of the imperial state terms into the negotiations:

> There still are quite close relationships between the officials of Venezuela and the United States, and I don't anticipate that the nationalization process or even our differences over monetary and world issues are seriously interfering with extremely good trade relations.[102]

Furthermore, the future capacity of the U.S. imperial state to disintegrate the Venezuelan state has not been limited in any significant way. Those ties to internal class allies within the government, state, and society that allow for the building up of support and permit the channeling of funds and so on continue unimpeded.

Certainly, the U.S. decision to bargain and negotiate with the Pérez government over the terms of Venezuelan dependence has not obscured the fact of persistent frictions over particular issues, such as oil prices, monetary relations, and trade arrangements. For example, the director of the Office of North Coast Affairs (which includes Venezuela) in the State Department, Frank Devine, stated in congressional testimony during May 1975 that "we have made known to them upon various occasions our lack of enthusiasm for their [oil] price increases . . . we did on various occasions make known to them the fact that we felt certain price increases were inappropriate and unhelpful to our economy and the world economy."[103] Nevertheless, the conflicts that do emerge are enveloped within and subsidiary to the larger common or mutual interests in that the U.S. dependence on Venezuelan oil is counterbalanced by the Venezuelan need for U.S. technology and markets. This interpretation of U.S.-Venezuelan relations is at the core of State Department thinking:

> [U.S.-Venezuelan] relations are basically good although there have been some sharp exchanges. President Pérez has had occasion to attack us—most recently on Secretary Kissinger's March [1975] proposal in Houston to establish a regional hemisphere agricultural center—and he attacked us vociferously on the Trade Act last year. He has given it to us on other issues too. Within this context, basic relations are good. The U.S. government has criticized Venezuela's petroleum pricing policy, not necessarily directly, but through OPEC. Still, Venezuela and the United States have strong economic ties. The United States gets more oil (crude and products) from Venezuela than any other traditional source. . . . There is a definite mutual interest on the part of the U.S. and the Venezuelan governments to have an amicable settlement of the petroleum nationalization because we do need their oil and they need our markets and technology. The United States is interested in a long-term supply of oil from Venezuela.[104]

Foreign policy differences are also mediated within these larger, overarching interests:

> [U.S.-Venezuelan] relations are generally good, but the rather dramatic change that has taken place in their financial fortunes as a result of increased income from oil and their more activist international foreign policy, and particularly the active role that they have been playing in the Third World consciousness, has inevitably caused some frictions. But the basic relationship is really very good. They have gone out of their way to say constantly

that they intend to be a secure source of supply of oil to the United States. As far as the frictions that have developed, for example, the open letter that President Pérez published in the New York *Times* in response to remarks by Ford and Kissinger at the United Nations, there has been an effort on both sides to minimize that kind of thing as a source of friction. The unusual amount of high-level exchanges we have had generated have played a principal role. They are more vocal supporters of these Third World issues, but I don't think they are any less an ally. We have many mutual interests and they have gone out of their way to point out that they have no intention of cutting off the supplies of oil.[105]

In the thinking of State Department officials, Venezuela's commitment to Third-World economic and political nationalism is overshadowed by and of secondary importance to its ties to the United States. The Pérez regime's large-scale financial support for the promotion of capitalist development in the Third World is the most visible example of the way in which it orders its priorities.

State Department policy continually emphasizes the need to anchor U.S.-Venezuelan relations within a long-term profitability perspective:

Venezuela has a high standard of living, a high per capita income, and U.S. business seems to feel a good and growing market, and that with a certain amount of goodwill and flexibility it may begin to grow and prosper ... the United States still sees the Venezuelan economy and market as a very attractive one ... [petrodollars] are greatly increasing the dimensions of the Venezuelan market for U.S. goods and services.[106]

Expanded future economic ties on the level of trade, markets, technology, and investments dictate negotiation and lead State Department policy makers to compare Venezuela favorably with Brazil and Mexico, which have both achieved relatively high growth rates based on foreign multinationals and forms of capitalist modernization from above with considerable state involvement: "We are developing an approach in which we think of Venezuela, Brazil, and Mexico as countries with more rapid developmental potential."[107]

The State Department has consistently emphasized the importance of economic considerations in its evolving response to the Pérez government in Venezuela. This is most clearly evident in its opposition to Venezuela's exclusion (because of the country's OPEC membership) from the generalized system of tariffs contained in the 1974 U.S. Trade Act. In congressional testimony, State Department officials have minimized Venezuela's role in "shaping the strategy and game plan of OPEC"[108] and counterposed the long-term possibilities for private capital accumulation, the expanding market for U.S. exports, and the country's role as a reliable oil supplier, especially in times of global crises:

> I don't think anybody is happy with the actions of OPEC, but I think that we should look at those countries that did continue to supply us oil and hopefully not discriminate against them, particularly as we look forward to the future. Both Iran and Venezuela are primarily one product economies and we hope that their diversification will encourage exports of products from the United States and that they can afford to buy these products and use them in their economies.[109]

Secondarily, State Department officials point out that Venezuela's leadership position within OPEC on the oil price issue is ultimately of less consequence than its refusal to allow OPEC to be employed as a political weapon against the United States and the other advanced capitalist countries and in support of Third-World proposals for a global socioeconomic redistribution of wealth:

> OPEC is really a Venezuelan creature. Venezuela has constantly been out front in OPEC on certain issues. Certainly, they have constantly been for higher prices.... But today, and for the last one and a half years, we see them as an important middle, and on some issues, moderating influence, particularly with regard to the use of OPEC as a political weapon.[110]

State Department opposition to Venezuela's exclusion from the generalized system of tariffs is essentially rooted in long-term considerations. First, during the 1973 Arab oil embargo, Venezuela "went out of its way to continue the supply [of oil]."[111] Second, countries such as Nigeria, which "is an increasingly important supplier of crude petroleum to the United States," expressed "serious concern" over the congressional action.[112] Third, there was a uniformly hostile Latin American response to Venezuela's exclusion from the tariff preferences. Of course, these considerations befit an agency of the U.S. imperial state whose primary concern is with maintaining the U.S. position as a competitive imperial power globally, in comparison with locally anchored U.S. capital, which finds its most ardent political supporters within the U.S. Congress.

U.S. DEPARTMENT OF THE TREASURY

The overall position of the U.S. Department of the Treasury toward the national development strategy of the Pérez government in Venezuela is rooted in a number of historic institutional concerns incorporated in the notion of "basic United States economic and financial interests."[113] The progressive nationalist populist period from April to October 1974, characterized by Pérez's promulgation of new laws and decrees affecting foreign investment, efforts on the part of the government to contain prices and increase salaries, and selective nationalization of foreign companies, elicited a hostile response from Treasury officials. The primary source of this hostility was the "regrettable" decision to nationalize the U.S.-owned iron ore and petroleum industries. Expropriation, per se, is perceived by the Treasury Department as a "negative" event that inevitably has adverse affect on the opportunities for U.S. capital expansion in the particular country in which it occurs.[114]

In January 1975, a number of interrelated factors induced Treasury officials to rethink and reassess the department's position on Venezuela: the Venezuelan decision to adhere to the notion of "swift, adequate and effective" compensation in respect of the iron ore companies indicated a desire on the part of Pérez to maintain close and comprehensive relationships with foreign capital; a recognition of the bourgeois nature of the regime as embodied in the plans flowing from the nationalization projects being pursued, allowing for multiple areas for foreign and national private investment; and the incapacity of the U.S. government (through the Treasury Department in particular) to exert economic pressures on Venezuela from the outside in order to reverse the nationalist trajectory. The Treasury decision to support a policy of accommodation and negotiated conflict must be located in the context of Venezuela's nonvulnerability to external economic pressures combined with the decision of the Pérez government to promote, and offer increasing concessions to, foreign capital as part of an overall strategy to promote the growth and expansion of national bureaucratic and private capital in collaboration with state-regulated foreign capital.

The Treasury Department's movement away from an overriding concern with the nationalization issue has been accompanied by an increasing tendency to view Venezuela in terms of its global position and role within the world capitalist economy (and its impact on the U.S. position) and less as a regional actor: "Treasury has a tendency to view things globally and to view bilateral relations in a global context."[115] Operationally, this has meant a focus on Venezuela's contribution to the U.S. balance of payments, terms of trade, and terms of exchange. Treasury officials have also given increased priority to Venezuela's role within OPEC:

> The attitude here generally is that Venezuela has been one of the more aggressive countries in seeking price increases over time, and on the price issue it is one of the countries that has given us the most problems. The Venezuelans play a very smart game, at least in terms of short-run strategy. They stimulated the birth of OPEC with the precise purpose of accomplishing what they accomplished.[116]

Treasury hostility to Venezuela's role in OPEC, however, is counterbalanced by the decision of the Pérez government to recycle surplus oil revenues through the so-called international banks (approximately $1.54 billion in 1974) for redirection into semiperipheral areas of the world economy within the U.S. sphere of influence (specifically Latin America). These petrodollar funds became available at a time when the capacity of the United States to engage in such activities was limited. They not only had the effect of buttressing pro-U.S. regimes in the hemisphere but also of creating new profitable opportunities for U.S. exports and investments in the area. Close collaboration between Treasury-appointed U.S. officials in the World Bank and the Inter-American Development Bank and their Venezuelan counterparts contributed significantly to the pursuit of these "basic U.S. economic and financial interests":

> Our concern is with basic U.S. economic and financial interests.... Our interest in Venezuela over the past couple of years relates to oil prices and there are some negative feelings here at OPEC generally. At the same time, Treasury is working rather well with the Venezuelans in the particular context of the Inter-American Development Bank in encouraging them and finding ways in which they could usefully recycle oil revenues to Latin American countries through the Inter-American Development Bank.[117]

Conflicts that do emerge are over marginal issues with limited consequences for Venezuela. Treasury Department resistance to authorizing an increase in foreign military sales credits to Venezuela (over strong Defense Department objections) is a case in point:

> Our attitude is that here's a country with enormous reserves that can go out into the world market and buy for cash all reasonable military needs and why should the U.S. government provide credits. We see no trade or economic justification for that credit. It is a reflection of our assessment of Venezuela's financial situation which derives from the increase in oil prices, but not retaliation for increased oil prices.[118]

Treasury's policy of "conflict and collaboration" with Venezuela has, since the beginning of 1975, emphasized negotiation (for the "best deal") rather than confrontation politics. From the perspective of Treasury officials, there are at present no serious problems between the United States and the Pérez government that raise the issue of the continued maintenance of capitalism and the capitalist mode of production in Venezuela. The arena of conflict —limited nationalization, controls on foreign capital exploitation, oil price increases, foreign military sales credits, and so on—is negotiable precisely because it does not call into question the adequacy of the ongoing relationship. That is, no effort to take Venezuela out of the capitalist politicoeconomic orbit has been or is envisaged by the current regime. On the contrary, new areas of expansion for imperial capital have been opened up and the internal market for U.S. goods and services has (due primarily to the impact of oil revenues) increased rapidly. No changes have taken place of the sort that would rupture ongoing ties and sanction political confrontation.

The Treasury Department counterposes Pérez's radical rhetoric and expansive foreign policy to the substantively more important fact of Venezuela's accommodative posture toward the United States on a day-to-day basis in such key areas as aid, investment, supply, trade, technology, and management:

> The Venezuelans have taken a rather radical voice in the world forums and a position of Third World leadership. They will tweak our nose, but in terms of practical day-to-day relationships I don't think we have any serious problems and I don't contemplate any unless the oil negotiations fall apart.[119]

Venezuela's implementation of the Andean Pact provisions, particularly as they relate to foreign investment, have been characterized by Treasury officials as essentially pragmatic. In other words, the investment laws have been minimally enforced and coexist with new and expanding opportunities for U.S. capital investments:

> Businessmen in Venezuela are fairly uncertain. On the one hand, they are aching to do something there because it does look like a growth territory and profits can be made. On the other hand, Venezuelans do engage in a great deal of populist rhetoric against foreign investment generally. Investment laws are going through sharp modification—the adoption of the Andean Pact and so on—and not always with a clear view of how they are going to be administered. However, the Venezuelan government has generally been rather pragmatic in these matters.[120]

Even Venezuela's active support of increased oil prices is ultimately designed to promote national capitalist development from above in collaboration with foreign, mainly U.S., capital.

The types of structural changes engineered by the Pérez government, limited to specific areas of industrial growth, have been accompanied by increased opportunities for U.S. industrial investors. The sectoral nationalization strategy has undercut the possibilities for class transformation, while the national development policies have, in general, strengthened the position of the dominant capitalist classes in society. As a result, Treasury officials have responded favorably to the Venezuelan transformation: "On all issues but the petroleum price there is fairly general admiration for what the Pérez government has attempted to accomplish within Venezuela."[121]

The limited changes that have occurred have taken place within a private capitalist economy and have increased the incentives for private capital. In sharp contrast to the anticapitalist development strategy of the Allende government in Chile, which tied nationalization to a larger redistributive policy and attempted to transform property ownership, the Venezuelan experiment under Pérez is, in the words of one Treasury official, "unradical enough to be within the bounds of tolerance":

> The range of differences [between Chile and Venezuela] are enormous. There is generally a sympathy with the notion of income redistribution that does not destroy the private sector and the incentive to produce. There is a relative lack of sympathy to converting everything to the public sector. A strong private sector is important to development. The Treasury position is that increasing state ownership introduces increasing inefficiency into the development process. In Chile you had a situation in which a minority president attempted to exercise a mandate which was incompatible in that society. . . . Economically, he made a shambles of the place. The massive redistribution of income created an enormous rate of inflation and destroyed the incentives to invest. The Venezuelan thing is very different. There you have populism which is not revolutionary in the sense of the immediate reform of all government structures. They are working more piece-by-piece. Pérez is going about it so that it's unradical enough to be within the bounds of tolerance.[122]

"Sympathy," but not implementation of income redistribution, and a "strong private sector" are enough to convince even conservative Treasury that Pérez is tolerable—for U.S. business interests.

U.S. DEPARTMENT OF DEFENSE

The U.S. Department of Defense located Venezuela within a primarily regional context (Latin America) as distinct from other agencies of the U.S. imperial state: "It looks like Venezuela may be emerging as one of the leaders in the whole Caribbean, Central American, South American group of nations. They aspire to a position of leadership and are achieving a certain amount of it."[123] In assessing U.S.-Venezuelan relations, Pentagon officials invariably place a considerable emphasis on Venezuela's role as an oil supplier and on its strategic location within the Central America-Caribbean area:

> Venezuela is undoubtedly important as a source of oil and its proximity is important in times of national emergency. . . . Oil is the biggest thing and its strategic location is important vis-a-vis the Panama Canal and other areas of the Caribbean. A friendly neighbor is important to any future strategic consideration involving that area of the world, for example, the question of submarines.[124]

Defense officials assume a relatively sophisticated stance in discussions regarding the nationalization policies of the Pérez government. First, they distinguish between the limited capacity of the U.S. government to realize politicoeconomic measures to reverse the nationalization process and the type of nationalization that has occurred—limited, sectoral, and within a capitalist context—which, under the circumstances, is perceived as the only alternative to a more profound transformation, whether statist or socialist:

> [Despite the nationalization policy] we would assume that the U.S. companies would continue to operate. But what really can you do? If they want to nationalize they are going to nationalize. All you can hope for is just compensation.[125]

Second, reflecting their particular institutional concern, Pérez' policies are not viewed by Defense Department policy makers as representing a threat to existing military ties. On the contrary, limited nationalization is quite compatible with the ongoing relationship between the U.S. and Venezuelan armed forces:

> There has been no real change or difference in the level of interaction between the U.S. and Venezuelan military since Pérez. It's an elected government. The military has their position and I don't think that, as governments go in Venezuela, there has been much significance for U.S. military and host country military relations. They have become a relatively sophisticated nation militarily and they have a desire to upgrade the type of assistance—from technical assistance to assistance at the joint staff and headquarters level. Except for the fact that they have become an advanced and sophisticated nation militarily, I don't think there has been any big changes in our relationship. I would say it's a very good relationship.[126]

The resultant Department of Defense policy is one that adheres to the position that it is imperative to maintain the existing channels of influence with Venezuela (political, economic, and military) and to negotiate what are essentially marginal conflicts that exist between the two countries. Rather than risk a fundamental rupture of relations as the result of a shortsighted policy, such as the application of sanctions because of the country's membership in OPEC, Venezuela's role as a strategic resource supplier (oil) and its geopolitical position within the hemisphere dictate negotiated settlements of disputes:

> There is a feeling in certain segments of our government that OPEC countries should be excluded from preferential [tariff] treatment—which we don't agree to in the case of Venezuela. I don't approach it from the role of Venezuela in OPEC but from the role of strategic location and a very

> strategic oil supplier, and we should be careful not to build any ill-will. Very careful. And we feel that OPEC is their business and we don't feel we should discriminate against them in any other areas because of that.[127]

Because Venezuela's support of increased oil prices has not been accompanied by any negative spillover impact in terms of oil supplies to the United States or changing international alignments, Defense Department officials argue that confrontation over the oil price issue is counterproductive to long-term U.S. interests in Venezuela: access to petroleum supplies, collaboration in the international and hemispheric arenas on fundamental issues affecting the struggle between capitalist and anticapitalist forces and the maintenance of ties with key sectors of the state apparatus (for example, the military) to allow for the possibility of overthrowing a future anticapitalist (socialist) regime that may emerge in a situation where social revolutionaries gain control of the Venezuelan government.

An important indicator of Defense Department attitudes may be seen in the support given Venezuela's request for increased foreign military sales credits over objections from other agencies of the U.S. imperial state, in particular, the Treasury Department:

> Venezuela is important from a national strategic point of view. Of course it makes it very important that we maintain good military-to-military relations. Venezuela has not been a grant material country. We do have a problem now with Venezuela. They are very eager for FMS [foreign military sales] credits and they have been one of the most active Latin American countries in seeking and planning for it. On the other hand, recent attitudes in the United States, that is, in the Department of the Treasury, have been to take the attitude that Venezuela doesn't need any more FMS credits and we shouldn't give them to OPEC countries. . . . Treasury, parts of the State Department, and various agencies have one sort of feeling about Venezuela: certain sorts of benefits should obtain and they shouldn't get anything special. Then there are others who realize the value of Venezuela, as we in Defense do. We desire to be as forthcoming as possible with Venezuela. We do not desire to limit or restrict anything to them due to their being an OPEC nation. We feel that we would lose more than we would gain.[128]

Declining foreign military sales credits are equated by the Pentagon with a parallel decline in follow-up servicing requirements and, hence, a weakened U.S. influence (leverage) within the Venezuelan armed forces. As one Pentagon official noted:

> What happens then, for instance, we are not able to offer the FMS credits. If they don't get it from us they have three options: buy on cash from the U.S., or buy on cash or credit from third countries. If we drive them away

from American military equipment by not providing credits, then pretty soon spare parts, repair equipment will decline, then technical assistance; they need to continue to maintain it but it will not be coming from the United States but from third country sources, and so we will further lose our influence there.[129]

It is precisely the weakening of ties with such critical sectors of the military apparatus that Department of Defense policy makers view as detrimental to long-term U.S. interests in Venezuela and, given the latter's regional position, within Latin America as well.

However, the Defense Department emphasis on the supply question is secondary to its stress on the importance of expanding political ties with the Venezuelan military. A recent study of U.S. Military Assistance Advisory Groups around the world, including Venezuela, concluded that "except for time spent on internal administration, MAAG efforts are devoted heavily to foreign military sales and dialogue, particularly in countries no longer receiving grant material assistance."[130] According to U.S. military personnel assigned to the Military Assistance Advisory Group in Venezuela, approximately half of their work time during 1974 was spent in "dialogue" activities:

Dialogue consists of *influence,* representation, and information exchange. *Influence is generally described as a means to develop rapport so that the host military will more readily accept suggested improvements and the American way;* representation is generally regarded as a protocol function to establish or enhance working relationships; and information exchange swaps ideas and/or information with host military officials on matters not necessarily related to one's occupational specialty (emphasis added).[131]

U.S. military personnel are heavily involved in creating important liaison groups within the Venezuelan military, preparing for any political eventuality. The primary ties established through "influence," "representation," and "information" exchange are to be thought of as strategic gains allowing the United States access points at critical crises moments that sooner or later will emerge in the course of Venezuela's capitalist development. The short-term exigencies of Treasury and State, hence, conflict with the long-term views of the Pentagon, a struggle over the immediate and long-term interests of American capital.

The Executive agencies of the U.S. imperial state involved in the elaboration and implementation of U.S. foreign policy function within a common politicoeconomic framework. In the case of Venezuela, within each agency a primary focus on those political and economic factors designed to facilitate the maintenance of the ongoing conditions for capitalist accumulation and expansion has been observed. The tactics and evolution of each agency's position must be located within the context of these overlapping interests. But the repetition of these basic concerns has also been paralleled by specific variations in positions taken by particular agencies, at different periods, reflecting their specific constituencies and unique institutional concerns.

U.S. CONGRESS

While the Executive branch of the U.S. government and key sectors of the U.S. capitalist class have opted for a negotiated resolution of existing disputes between the United States and Venezuela, the U.S. Congress has assumed a uniformly hostile posture. This negative congressional response is not directed toward the development-nationalization policies of the Pérez regime but rather to Venezuela's membership in OPEC and Venezuela's role in it. As one congressional figure noted: "On the organization of the [OPEC] cartel and the unilateral increase in oil prices, Venezuela has played a leading inspirational role."[132]

The minimal congressional response to the nationalization of U.S. property interests in Venezuela has been dictated, in large part, by the payment of compensation:

> There are no feelings in Congress about the nationalization policies of the Pérez regime. That issue has been fading generally and really is only raised seriously if a question arises of no compensation or inadequate compensation.... The nationalization issue [as regards Venezuela] is a nonissue. No oil company has come around here and complained.[133]

The willingness of the Venezuelan government to compensate satisfactorily expropriated U.S. investor interests eliminated the need to invoke legislative sanctions (or the threat of them) and reflected the fact that the new development project was not moving in the direction of social transformation.

Unlike Chile under Allende, Pérez' Venezuela is not perceived as a nationalist pivot in Latin America. There are probably two primary (interrelated) reasons for this visible lack of concern over the policies being pursued by the current regime. First, the type of nationalism being promoted is inconsequential in that it does not conflict with substantial foreign capital expansion and it is rooted in private capital expansion. Even on the hemispheric level, this orientation is in evidence. In the recently formed Sistema Económico Latinamericano (SELA), "the battle lines have [already] been drawn between the Cuban advocates of state participation [if not state control] and the Venezuelan supporters of private enterprise."[134] Second, since Pérez's nationalism is tied to capitalist development, it is perceived as less of a problem than the economic consequence of the equalization of exchange resulting from oil price increases, which more directly affects U.S. oil consumers than the political challenge of limited nationalization.

The U.S. Congress views Venezuela "in global terms to the extent that there is concern expressed in terms of global energy supplies [trade bill, OPEC, and so on] rather than as a Latin American country" and it is in this context that the basis of congressional hostility must be sought. Congressional antagonism is rooted in Venezuela's role within OPEC and the efforts of OPEC to change the global bargaining equation, to equalize the terms of exchange between the industrialized nations and the Third-World countries:

> Congress doesn't look at Venezuela as a whole at all. Who is taking an interest depends on the particular issues involving Venezuela. If the issue is the oil price increase then the interest is fairly widespread. At the moment there is a lot of hostility toward Venezuela in Congress; a lot of the basis is oil, a belief that a country we view as a friend took an unfriendly action which hurt. No distinction is made between OPEC as a unit and the organization of its individual members.[135]

This position was most forcefully stated during hearings on generalized tariff preferences held by the Subcommittee on Trade of the House Ways and Means Committee in May 1975. One congressional staff consultant described the climate within the subcommittee just prior to debate on the bill: "Hostility to Venezuela is located absolutely everyplace [in Congress]. When chairman [William J.] Green introduced the bill . . . he was shocked at the [hostile] response of the subcommittee members [towards Venezuela and OPEC]."[136]

In the course of the subsequent debate, efforts by State Department officials to justify Venezuela's inclusion in the generalized tariff preferences on the grounds of its refusal to support the 1973 Arab oil embargo made no impact on subcommittee members. The following exchange between Robert S. Ingersoll, deputy secretary of State, and Representative Gibbons is illustrative of this point:

> Ingersoll: Well, Venezuela and these countries that we have mentioned continued to supply oil to us during the embargo by the Arab OPEC countries.
>
> Gibbons: They either had to supply it or eat it, one of the two. With Arabs not supplying us there must have been a glut of oil in Arab countries seeking to go to other places. Really, Venezuela and some of these other countries were just serving their own self-interest.[137]

Later in the congressional debate, Venezuela's motives in supplying the United States with oil during the 1973 embargo were again soundly denounced in terms of economic self-interest, while its active role in support of oil price increases was implicitly criticized because of the alleged impact of these increases on the U.S. economy. Representative Archer described the feelings of many congressmen when he stated:

> A major portion of our recession today has been precipitated by the quadrupling of the price of oil. You are giving these countries a pat on the back because they did not participate in the embargo but it was to their selfish interest not to participate in the embargo. Many of them were getting $16 to $20 a barrel for oil during that period of stress in this country. . . . They did not continue to supply us because they had concern for us. They were helping themselves far more than they were helping us. I don't think it is appropriate to tell this committee today that we should be forever grateful because they jacked the prices up to $16 or $20 a barrel and got that price for oil at a time when we were, in the vernacular, "over the barrel."[138]

The overall attitude of the subcommittee toward Venezuela was summed up by Representative Mikva: "With Ecuador and Venezuela as friends we don't need many enemies. I think that is the attitude that is being reflected."[139]

U.S. congressmen are elected on the basis of multiclass or class collaborationist positions that unite national local capitalists and U.S. workers. They may perceive the basis of this coalition as rooted in access to cheap raw materials that help to sustain a high standard of living. The actions of OPEC under Venezuela's inspirational leadership are thus perceived as undermining the social basis of their political power within the country. Thus, the issue for Congress is essentially anchored in a coalition of local interests, reflecting the desire of the Legislative branch of the U.S. government to unite all internal classes against foreign (strategic) raw material producers in order to maintain the U.S. standard of living at the expense of the peripheral and semiperipheral areas of the world economy. This hostility, not to nationalization, per se, but to the equalization of the terms of exchange through oil price increases, is captured in the blunt observation of a senior congressional staff consultant:

> U.S.-Venezuelan relations boil down to what the hell the price of oil is from the perspective of the Hill. . . . Congress sees Venezuela in oil terms only. Oil completely overshadows any other relationships. It's not a foreign policy problem but a domestic problem.[140]

Whereas the Executive branch of the U.S. government plays an important role in support, and is always directly accessible to, imperial capital, the U.S. Congress tends to be more of a sounding board for local capitalists. This locally anchored imperialism vis-a-vis the Third World parallels an always aggressive reaction on the part of the legislature to interimperial rivalries (vis-a-vis Europe and Japan) because the costs and scarcities stemming from such rivalries threaten to undermine the economic cushions that allow the local bourgeoisie its preeminent position over the wage and salaried classes.

INTERNATIONAL BANKS

The transformation of Venezuela's position within the world capitalist system—from a dependent to a more autonomous capitalist economy with regional expansionist aspirations—became evident by the early 1970s and was largely rooted in the accumulated wealth derived from oil revenues (petrodollars). Venezuela's enhanced standing within such international capitalist financial agencies as the World Bank, the International Monetary Fund, and the Inter-American Development Bank is a direct consequence of its enormous financial resources, which have allowed it to move from being a recipient of credits to a donor of funds to these institutions.

By the end of 1974, the Pérez government had disbursed some $700 million of an estimated $2.5 billion in oil revenues that it had agreed to recycle through a number of bilateral and multilateral lending agreements.[141] The multilateral loan commitments during 1974 included $500 million to the World Bank, $540 million to the International Monetary Fund oil facility, and $500 million to the Inter-American Development Bank. The most important bilateral arrangements are with the Central American republics and will amount to between $600 million and $700 million over the next five years. "These arrangements include a facility to compensate the Central American republics for the increased cost of oil imports for domestic consumption, an arrangement to help stabilize international coffee prices, and a subscription to bonds of the Central American Bank for Economic Integration."[142] During 1975, Venezuela also committed $60 million to the Andean Development Corporation and $25 million to the Caribbean Development Bank. In March, a loan of approximately $300 million was made to the International Monetary Fund oil facility, a short-term loan fund designed to aid member countries experiencing balance of payments difficulties stemming from rising oil prices; and, according to *Business Week* "another $1.1 billion is earmarked for agencies such as the Inter-American Development and the Ande[an] Development Corp[oration]."[143]

The transformation of Venezuela from a net borrower to a net lender of capital served to cement the already close relationship that existed between it and the international financial institutions. One World Bank official discoursed on the nature of that organization's historic ties with Venezuela:

Relations are very good. Venezuela has lent to the Bank x million dollars which will be used to finance projects in the less developed countries. They have viewed the World Bank as a channel for recycling some of the petrodollars. The Bank has taken a very positive and enlightened attitude toward Venezuela. Generally, our relations with them have traditionally been good. Traditionally, we have been involved in very successful policies in Venezuela and, therefore, our image is very good.[144]

The international banks have historically functioned to promote U.S. economic interests and specifically to promote U.S. overseas investment and exports. In most instances, U.S. personnel or appointees favorable to U.S. interests occupy key positions in the banks or have sufficient voting power to influence loan allocations, for example, for or against a regime, as was illustrated in the cases of the Velasco government in Peru (1969–71), the Allende government in Chile (1970–73), and others.[145]

The ability of the World Bank and the Inter-American Development Bank to gain access to Venezuelan petrodollars, accompanied by increased Venezuelan leverage incommensurate with the amount of its contribution, has been beneficial to the banks in a variety of ways. First, Venezuela no longer requires loans or credits from these organizations. On the contrary, instead of making demands on bank resources, it has become a provider of large-scale funds for Bank enterprises. A senior World Bank economist stated that "since the Pérez government there has been a great improvement in relations, the main reason being that we have changed our position from being a net lender to a net borrower."[146] Second, the decision by the Pérez government to channel Venezuelan financial resources to these U.S.-influenced international agencies for subsequent redirection into the peripheral areas of the world economy has served a double purpose. It has occurred at a period when both the World Bank and the Inter-American Development Bank appear to be experiencing problems in raising sufficient financial capital in the United States to carry through their projected policies and it has served to increase the leverage of these institutions in those areas, principally Latin America, where Venezuelan funds are largely being channeled. The efforts of the international banks to stabilize pro-capitalist regimes and promote capitalist development in these areas have been greatly enhanced by the actions of the Pérez government. Venezuela's contributions to the Inter-American Development Bank have clearly played an important part in stabilizing the financial resources of the Bank:

We placed a bond issue beginning in 1974. It was not a large amount but Venezuela was the first Latin American country where we were able to raise resources on a long-term basis . . . at the present time a replenishment of the resources of the Bank is being considered—an increase in ordinary capital resources and an increase in "soft" window resources. Venezuela has been willing to make a special contribution to ordinary capital and to soft resources, and has also agreed that some bolivars that we have be used on a convertible basis. They are making fully convertible resources, which is something that goes beyond what other Latin American countries are doing.

All this indicates that Venezuela, through the Bank, has been channeling important resources to the development of the region. We are mobilizing in Venezuela freely convertible resources.[147]

The World Bank has also benefited from Venezuela's financial support:

> The World Bank relationship with Venezuela is very cordial. We have obtained $500 million at 8 percent from Venezuela at a time when the World Bank was trying to get reserves to lend to other countries for development projects.[148]

Third, despite Venezuela's present position as a nonborrower of capital, the influence of the World Bank, in particular, in the implementation of various development undertakings remains highly visible. World Bank technical advisers, operating through the pro-capitalist elements that control the Venezuelan Investment Fund, continue to advise the Pérez government on long-term development projects:

> ... the World Bank has become a kind of technical adviser to the Venezuelan Investment Fund, which is the entity that is going to take care of the surplus resources that Venezuela has with the accession of the oil price boom ... the World Bank has become an assistant for the function of domestic investment of resources, especially in advising the government on long-term development projects. So there is an agreement between the World Bank and the Venezuelan government in the sense that we give advice as to the quality of the projects they are involved in. We appraise projects. Technical assistance of this sort is very innovative in terms of the World Bank's policy. Usually the World Bank does not do any technical assistance except as part of lending projects, but in the case of Venezuela we have been doing the technical assistance without having been in on the actual process of lending —which is innovative. The Bank hasn't done this in the past.[149]

This close collaboration between World Bank officials and the technocratic sector of the Venezuelan national bourgeoisie reflects an implicit recognition and approval of the government's development program.

The main beneficiary of Venezuelan funds is neither the banks nor the loan recipients or their donors but U.S. capital, which will thus receive a new impetus for exports, new areas for investment, and further expansion of capital markets to which it has access. Venezuelan donations of capital to institutions in which there is no commensurate increase in decision-making power suggests that the capital inflows are strengthening the power of the United States, which does wield power within the banks. Thus, the increasing oil revenues accruing to Venezuela, far from weakening the United States, lead to an extension of its power precisely at a moment when it appeared to be in a declining position.

Officials of the World Bank and the Inter-American Development Bank maintain a neutral posture in assessing the nationalization policies of the Pérez government: "Our own position is very technical," observed a World Bank official;[150] "the Bank does not enter into areas which are subject to the political decision of each country," a high-ranking Inter-American Development Bank official maintained.[151] There is also a pronounced tendency to resort to technocratic rhetoric in any discussions of Pérez overall development strategy. Any adequate explanation of these positions must take account of at least two important factors. First, Venezuela's decision to pay compensation to the former U.S. owners of the nationalized iron ore and petroleum industries has led to the designation of the government as a "creditworthy" one.[152] The position of the World Bank is quite explicit on this point: "We don't have a position on nationalization. Our board has ruled that provided there is adequate compensation we don't care if nationalization takes place or not."[153] Second, because oil revenues have eliminated the possibility of external financial dependence, the banks have realized that they lack the economic levers with which to exert pressure on Venezuela (as they did to Chile during the Allende period). A World Bank official alluded to this new factor in the equation:

> In summary, [the World Bank and Venezuela have enjoyed] a very close, very fruitful relationship with a lot of cooperation coming from Venezuela. The only difference between now and before is that Venezuela is in a completely different financial position. With the increasing price of oil, Venezuela has resources that it did not have before. That is the change, the modification.[154]

The international banks support and justify the Pérez government's decision to export capital resources by reference to the country's supposed lack of absorptive capacity at the present time, its incapacity to absorb these new funds without generating severe inflationary pressures. World Bank officials term this lack of absorptive capacity a "technical" problem:

> The problem there is the problem of the capacity to absorb the surplus quickly. You may have financial resources, but the key needs of development itself take time. What are the bottlenecks that determine the capacity to absorb? One is human resources; development requires people for big development tasks. It takes time to get projects going, invest money and train

people abroad. The advantages that accrue to sending the surplus out are: (1) internally it would be inflationary; (2) [it] wouldn't be [possible] to undertake the projects so quickly; (3) therefore, [there is the] need to use the money outside and get interest on it. You also have to make projects, do careful studies on the projects, which takes time. Finally, there are monetary limits—tremendous expansion in the short run would generate inflationary pressures which would be intolerable.[155]

Unable to absorb all its new financial resources, given the existing class structure, Venezuela has tied itself to the projects and perspectives of these international institutions, collaborating in the export of capital to other proimperial ruling classes in the periphery in order to finance private capital expansion and maintain conditions for capital accumulation.

Within the World Bank, hostility toward Venezuela's role in OPEC (essentially its advocacy of increased oil prices) is muted at the present time, but the possibilities exist that this issue could exacerbate international rivalries among bank members and become a future source of conflict between the World Bank and Venezuela:

At the moment, the World Bank and Venezuela are very close. The only problem that may arise is this: in the Bank are many countries being hurt by the petroleum price increase and, therefore, it seems to be a kind of reaction against some of the countries in the World Bank to take a position against oil-producing countries—and that could eventually be a source of problems.... The World Bank has to be extremely hesitant to pass judgement on questions which are in the public controversy right now—particularly when related to questions of OPEC countries where the board could be split on lending.[156]

Thus, the fact that the United States is benefiting from the increased oil revenue accruing to Venezuela while other capitalist countries are being adversely affected suggests that in the new world alignment Venezuela will be an important ally of the United States—jointly exploiting others and negotiating new terms of association.

NOTES

1. Henry Kissinger, "Strengthening the World Economic Structure," speech presented at Kansas City, Mo., May 13, 1975 (U.S. Department of State, Office of Media Services, Bureau of Public Affairs).
2. Ibid.
3. "Kissinger on Oil, Food, and Trade," interview in *Business Week,* January 13, 1975, p. 66.
4. Richard S. Frank, "Kissinger Stays Up Front," *National Journal,* June 21, 1975, p. 922.

5. U.S., Congress, Subcommittee on Energy of the Joint Economic Committee, *U.S. Foreign Energy Policy,* 94th Congress, 1st sess., September 17 and 19, 1975, p. 53.

6. See U.S., Congress, Senate, Joint Committee on Atomic Energy, *Towards Project Interdependence: Energy in the Coming Decade,* 94th Cong., 1st sess. Joint Committee Print, Congressional Research Service of the Library of Congress, December 1975, p. vii.

7. U. S., Congress, House, Committee on International Relations, Special Subcommittee on Investigations, *Oil Fields as Military Objectives: A Feasibility Study,* 94th Cong., 1st sess., Committee Print, Congressional Research Service of the Library of Congress, August 21, 1975, p. 5. For figures on oil imports as a percentage of total domestic requirements in Western Europe, Japan, and the United States, see ibid., Table 5, p. 7.

8. Kissinger, op. cit.

9. Ibid.

10. Interview 2-7, U.S. Department of State, Washington, D.C., August 4, 1975.

11. Thomas C. Enders, Assistant Secretary of State for Economic and Business Affairs, *An Action Program for World Investment,* September 5 and 6, 1975, U.S. Department of State Publication 8780, General Foreign Policy Ser. 289, p. 12.

12. Interview 2-7, U. S. Department of State, Washington, D.C., August 4, 1975.

13. Interview 2-12, U.S. Department of State, Washington, D.C., August 7, 1975.

14. Ibid.

15. State Department official, quoted in Robert M. Smith, "Ambassador Urged U.S. Take Role in Venezuelan Oil Talks," New York *Times,* June 30, 1975, p. 45.

16. Ibid.

17. Interview 2-17, U.S. Department of the Treasury, Washington, D.C., August 20, 1975.

18. Interview 2-6, World Bank, Washington, D.C., August 1, 1975.

19. Interview 2-12, U.S. Department of State, Washington, D.C., August, 7, 1975.

20. See Norman Gall, "The Challenge of Venezuelan Oil," *Foreign Policy,* no. 18, Spring 1975, pp. 55, 56.

21. Edward Cowan, "4 Top Concerns Raise Gas Prices to 3¢ a Gallon," New York *Times,* July 2, 1975, p. 46.

22. U.S., Congress, Subcommittee on Energy of the Joint Economic Committee, op. cit.

23. See "Venezuela: Ford Weighs In," *Latin America,* September 26, 1975, pp. 298, 300.

24. See, for example, Henry Kissinger, "The United States and Latin America: The New Opportunity," speech presented at Houston, Tex., March 1, 1975, and "American Unity and the National Interest," speech presented at Birmingham, Ala., August 14, 1975 (U.S. Department of State, Office of Media Services, Bureau of Public Affairs).

25. Adviser to President Pérez quoted in "Caracas Shares Its Oil Wealth," Washington *Post,* December 28, 1974, p. A8.

26. "The Venezuelan Views," New York *Times,* January 26, 1975, p. 75. See also "Letter of January 7, 1975 from Carlos Andrés Pérez, President of Venezuela, to all of the Latin American Chiefs of State," reprinted in Organization of American States, General Assembly, *Special Report of the Action Taken by the Permanent Council and Background Relevant to the United States Foreign Trade Act of 1974,* OEA/Ser.P, AG/doc. 544/75, April 9, 1975, Fifth Regular Session, May 8, 1975, Washington, D.C., pp. 345-347.

27. Quoted in "Venezuela: Prudent Audacity," *Latin America,* February 28, 1975, p. 65.

28. Ibid.

29. Kissinger, "Strengthening the World Economic Structure," op. cit.

30. Interview 2-7, U.S. Department of State, Washington, D.C., August 4, 1975.

31. Ibid.

32. Interview 2-8, U.S. Department of Defense, Virginia, August 5, 1975.

33. Telephone conversation, U.S. Department of Treasury official, Washington, D.C., August 7, 1975.

34. Interview 2-9, U.S. Chamber of Commerce, Washington, D.C., August 6, 1975. See also U.S. Chamber of Commerce, *The Climate for Investment Abroad,* September 1974, pp. 102-03; investment figures are taken from U.S., Congress, Senate, Committee on Foreign Relations, Subcommittee on Western Hemisphere Affairs, *U.S. Relations with Latin America,* 94th Cong., 1st sess., February 21, 26, 27, and 28, 1975, p. 209.

35. "1974 Is Better Than Expected for Most Venezuelan Firms," *Business Latin America,* February 5, 1975, pp. 44-45.

36. "Venezuela's AD Administration Makes Ambitious Pledges," *Business Latin America,* March 20, 1974, p. 91.

37. Quoted in Business International Corporation, *Background Paper,* Roundtable with Government of Venezuela, Caracas, November 10-14, 1974, p. 52.

38. "What Foreign Firms Can and Cannot Do in Venezuela," *Business International,* January 3, 1975, p. 7.

39. Business International Corporation, op. cit., pp. 31, 50.

40. Ibid., p. 30. See also "Decision 24 in Venezuela: Harsher Than Anticipated," *Business Latin America,* May 29, 1974, pp. 175-76.

41. Business International Corporation, op. cit., p. 14.

42. Interview 1-1, Business International Corporation, New York, March 1975.

43. Business International Corporation, op. cit., p. 1.

44. "Venezuela Completes Iron Mines Takeover," *Latin America Economic Report.* January 17, 1976, p. 9.

45. Economist Intelligence Unit, *Quarterly Economic Review of Venezuela,* no. 1, January 1975, p. 10.

46. Interview 2-9, U.S. Chamber of Commerce, Washington, D.C., August 6, 1975.

47. Interview 1-1, Business International Corporation, New York, March 1975.

48. Interview 2-5, U.S. Department of State, Washington, D.C., July 31, 1975.

49. "What Foreign Firms Can and Cannot Do in Venezuela," op. cit., p. 7.

50. "1974 Is Better Than Expected for Most Venezuelan Firms," op. cit., p. 44; Business International Corporation, op. cit., p. 14.

51. Business International Corporation, op. cit., p. 4.

52. "What the Third World Wants: Interview with Venezuela's President Carlos Andrés Pérez," *Business Week,* October 13, 1975, p. 57.

53. Business International Corporation, op. cit., p. 52.

54. Council of the Americas, *The Trade Act of 1974,* Summary of COA Member Company Views on Hemispheric Trade and Relations, March 11, 1975, p. 6.

55. U.S., Congress, House, Committee on Ways and Means, Subcommittee on Trade, Generalized Tariff Preferences, 94th Cong., 1st sess., May 13 and 19, 1975, p. 49.

56. U.S., Congress, Senate, Committee on Foreign Relations, op. cit., p. 209.

57. U.S., Foreign Service and Department of State, "Foreign Economic Trends and Their Implications for the United States—Venezuela," June 1975, p. 8.

58. "How Firms in Venezuela Are Faring with Profits," *Business Latin America,* April 7, 1976, p. 107.

59. Interview 2-1, Exxon Corporation, New York, June 23, 1975.

60. Interview 2-3, Asiatic Petroleum, New York, July 14, 1975.

61. Ibid.

62. Interview 2-1, Exxon Corporation, New York, June 23, 1975.

63. Ibid.

64. "Venezuela: Oil Storm," *Latin America,* August 30, 1974, p. 266.

65. Interview 2-2, Mobil Oil Corporation, New York, June 24, 1975.

66. Interview 2-3, Asiatic Petroleum, New York, July 14, 1975.

67. Interview 2-1, Exxon Corporation, New York, June 23, 1975.

68. "Venezuela: Oil Storm," op. cit., p. 266.
69. Interview 2-2, Mobil Oil Corporation, New York, June 24, 1975.
70. Interview 2-1, Exxon Corporation, New York, June 23, 1975.
71. "Venezuela's Takeover of Oil Leaves Room for Foreign Firms," *Business Latin America,* March 19, 1975, p. 92.
72. Economist Intelligence Unit, *Quarterly Economic Review of Venezuela,* no. 2, April 1975, p. 6.
73. Interview 2-3, Asiatic Petroleum, New York, July 14, 1975.
74. Interview 2-2, Mobil Oil Corporation, New York, June 24, 1975.
75. Ibid.
76. Ibid.
77. Ibid.
78. Interview 2-3, Asiatic Petroleum, New York, July 14, 1975.
79. Ibid.
80. See "Venezuelan Compensation Is Accepted," New York *Times,* October 29, 1975, p. 36; "Venezuelan Oil Takeover Still Faces Several Hard Tests," *Business Latin America,* November 5, 1975, pp. 353-54; " 'Good Manners' in Oil Nationalization," *Latin American Week* (Buenos Aires), November 21, 1975, p. 7.
81. See Robert S. Smith, "Hammer Accused of Foreign Bribes," New York *Times,* October 9, 1975, pp. 1, 72. See "Venezuelan Panel Assails Occidental, Urges No Compensation for Its Holdings," *Wall Street Journal,* June 4, 1976, p. 7. A Venezuelan congressional commission appointed to look into these charges concluded, following a seven-month investigation, that Occidental had engaged in "irregular activities" in its efforts to gain the contracts. The commission recommended that the oil company not be paid compensation for its nationalized holdings.
82. "Venezuela's Compensation to Be Paid Mainly in Oil," *Latin American Economic Report,* December 19, 1975, p. 199. See also "Venezuela's Oil Industry Is Formally Nationalized," New York *Times,* January 8, 1976, pp. 35, 37.
83. "Venezuela's Compensation to Be Paid Mainly in Oil," op. cit., p. 199.
84. Joanne Omang, "Venezuelan Oil: Troubled Bonanza," Washington *Post,* December 28, 1975, p. A23.
85. Ibid. See also "Exxon Reaches Venezuelan Pact on Oil Purchases," *Wall Street Journal,* January 7, 1976, p. 2.
86. Interview 2-2, Mobil Oil Corporation, New York, June 24, 1975.
87. Franklin Tugwell, *The Politics of Oil in Venezuela* (Stanford, Calif.: Stanford University Press, 1975), pp. 77-80, 179-80.
88. Ibid., p. 134.
89. Ibid., pp. 139-40.
90. Interview 2-1, Exxon Corporation, New York, June 23, 1975.
91. Interview 2-12, U.S. Department of State, Washington, D.C., August 7, 1975.
92. Interview 2-4, U.S. Department of State, Washington, D.C., July 30, 1975.
93. Interview 2-7, U.S. Department of State, Washington, D.C., August 7, 1975.
94. Interview 2-12, U.S. Department of State, Washington, D.C., August 7, 1975.
95. Interview 2-7, U.S. Department of State, Washington, D.C., August 4, 1975.
96. Ibid.
97. Ibid.
98. Interview 2-5, U.S. Department of State, Washington, D.C., July 31, 1975.
99. Ibid.
100. Interview 2-7, U.S. Department of State, Washington, D.C., August 4, 1975.
101. Interview 2-5, U.S. Department of State, Washington, D.C., July 31, 1975.
102. Interview 2-12, U.S. Department of State, Washington, D.C., August 7, 1975.

103. U.S., Congress, House, Committee on Interstate and Foreign Commerce, Subcommittee on Oversight and Investigations, *FEA Enforcement Policies,* 94th Cong., 1st sess., April 9, 11; May 6, 7, and 8, 1975, p. 399.
104. Interview 2-7, U.S. Department of State, Washington, D.C., August 4, 1975.
105. Interview 2-4, U.S. Department of State, Washington, D.C., July 30, 1975.
106. Interview 2-5, U.S. Department of State, Washington, D.C., July 31, 1975.
107. Interview 2-12, U.S. Department of State, Washington, D.C., August 7, 1975.
108. Interview 2-5, U.S. Department of State, Washington, D.C., July 31, 1975.
109. Robert S. Ingersoll, Deputy Secretary of State, U.S., Congress, House, Committee on Ways and Means, Subcommittee on Trade, op. cit., p. 14.
110. Interview 2-12, U.S. Department of State, Washington, D.C., August 7, 1975.
111. Interview 2-4, U.S. Department of State, Washington, D.C., July 30, 1975.
112. Ingersoll, op. cit., p. 11.
113. Interview 2-17, U.S. Treasury Department, Washington, D.C., August 20, 1975.
114. Ibid.
115. Ibid.
116. Ibid.
117. Ibid. Like their international bank counterparts, Treasury officials also emphasize Venezuela's limited capacity, in the short term, to absorb the large new infusions of petrodollars. See, for example, U.S., Department of the Treasury, Office of Assistant Secretary, Trade, Energy and Financial Resources Policy Coordination, *The Absorptive Capacity of the OPEC Countries,* September 5, 1975, especially pp. 30-31.
118. Ibid.
119. Ibid.
120. Ibid.
121. Ibid.
122. Ibid.
123. Interview 2-8, U.S. Department of Defense, Virginia, August 5, 1975.
124. Ibid.
125. Ibid.
126. Ibid.
127. Ibid.
128. Ibid.
129. Ibid.
130. U.S., General Accounting Office, Report to the Congress, *Assessment of Overseas Advisory Efforts of the U.S. Security Assistance Program,* ID-76-1, October 31, 1975, p. 10.
131. Ibid., pp. 10, 11.
132. Interview 2-14, U.S., Congress, Washington, D. C., August 17, 1975.
133. Interview 2-16, U.S., Congress, Washington, D.C., August 18, 1975. The fact that big business does not look to Congress for support on overseas problems may also be due to the concentration of power in the Executive branch. Congress is considered more as a forum within which grievances may be aired, publicity garnered, and opinions expressed.
134. "SELA Sets Its Sails," *Latin American Economic Report,* October 24, 1975, p. 167.
135. Interview 2-16, U.S., Congress, Washington, D.C., August 18, 1975.
136. Interview 2-15, U.S., Congress, Washington, D.C., August 18, 1975.
137. U.S., Congress, House, Committee on Ways and Means, Subcommittee on Trade, op. cit., p. 13
138. Ibid., pp. 20-21.
139. Ibid., p. 26.
140. Interview 2-14, U.S., Congress, Washington, D.C., August 17, 1975.

141. See "Venezuelan Payments Position Favorable; Oil Proceeds Channelled to Development," *IMF Survey,* June 9, 1975, p. 176; World Bank, *Annual Report,* 1975.

142. "Venezuelan Payments Position Favorable," op. cit., p. 176.

143. "Venezuela: A Study in Third World Strategy," *Business Week,* October 13, 1975, p. 59; see also Economist Intelligence Unit, *Quarterly Economic Review of Venezuela,* no. 2, April 1975, p. 5.

144. See James Petras and Morris Morley, *The United States and Chile: Imperialism and the Overthrow of the Allende Government* (New York: Monthly Review Press, 1975); U.S., Congress, House, Committee on Foreign Affairs, *The United States and the Multilateral Development Banks,* 93rd Cong., 2nd sess., March 1974, Committee Print, Congressional Research Service of the Library of Congress.

145. Interview 2-13, Inter-American Development Bank, Washington, D.C., August 13, 1975.

146. Interview 2-6, World Bank, Washington, D.C., August 1, 1975.

147. Interview 2-10, World Bank, Washington, D.C., August 6, 1975.

148. Interview 2-6, World Bank, Washington, D.C., August 1, 1975.

149. Ibid.

150. Interview 2-13, Inter-American Development Bank, Washington, D.C., August 13, 1975.

151. Ibid.

152. Interview 2-10, World Bank, Washington, D.C., August 10, 1975.

153. Interview 2-13, Inter-American Development Bank, Washington, D.C., August 13, 1975.

154. Interview 2-6, World Bank, Washington, D.C., August 1, 1975.

155. Ibid.

156. Interview 2-10, World Bank, Washington, D.C., August 10, 1975.

CHAPTER 4

SUMMARY

THE TRANSITION TO INDUSTRIAL CAPITALISM

Revolution, in many historic instances, has laid the groundwork for large-scale capitalist development by overturning feudal restrictions on industry and creating internal markets. Out of these changes industrial capital has emerged as the dominant force over and against the agrocommercial strata. Not infrequently, parliamentary governments served as the political vehicle through which the new social forces exercised power. The Venezuelan experience bears some similarities but also significant differences with this classic pattern. The process of change, over the long run, has been manipulated and directed from above; there have been few barricades, revolutionary crowds, or mass mobilizations accompanying the process. It has been the state oil revenues that have served to underwrite the expansion of Venezuelan industrialization in much the same way as bourgeois revolutions in earlier periods provided the impetus for capitalist expansion. The products of democratic-capitalist revolution—the growth of an internal market, the increasing social weight of finance groups, and the consolidation of the parliamentary form of rulership—are, in the case of Venezuela, promoted by petroleum wealth.

While the various bourgeois regimes at times have relaxed restrictions on the participation of revolutionary parties (in this context, socialist revolutionaries) within the national parliament, they have maintained a tight control over extraparliamentary activity. The absence of mass mobilization contributes to the relatively static nature of the social order. As no mass organizations have emerged to influence and shape policy effectively, the masses have been, and continue to be, spectators to politics—marginal recipients subject only to the bounty of election campaigns. The tradition of class struggle and the resultant class-anchored mass organizations that emerged from the European bourgeois transformations are still in the process of formation in Venezuela. Up to now the trade unions and other mass organizations have been dominated by lower middle income strata that have harnessed their efforts to achieve the goal of an industrial capitalist society. The lack of autonomous, self-directed working class organizations is one of the chief results of the bourgeois transformation from above.

Another aspect of this nonrevolutionary style of capitalist development is the nature of the formation of the finance capitalist class. The transition to an industrial social formulation has been made with relatively little bloodletting among the powerful social groups. The oil-rich Venezuelan state's various subsidies to the nation's ruling class and the latter's partnership with foreign capital have allowed the rulers to change the type of resources upon which their economic power was based. This has assured a considerable continuity in the dominance of the family-based oligarchy.

In some cases the path toward industrial capitalism involved the conversion of commercial and banking wealth into industrial assets and business services; in others the process was preceded by the transfer of agricultural wealth to real estate, construction, and speculative commercial activity. More recently, there has been a reciprocal flow toward large-scale commercial agriculture. The differences between nonindustrial and industrial capital have been almost obliterated. Today the diversification of the holdings of key finance groups minimize the political significance of sectoral conflicts. The process of industrial capitalist development, in the final analysis, was not hindered by the existence of a comprador bourgeoisie; on the contrary, when it became profitable, the latter adapted to conditions and began to shift investments from import-export activities to establishing manufacturing enterprises.

Because the emergence and consolidation of finance capital were in large part dependent on state intervention and support and integrated with foreign capital, a substantial professional-bureaucratic stratum developed. A portion of higher paid state employees came to press for greater control over the U.S.-owned oil enclave as a way of furthering economic expansion in general and state intervention and social mobility in particular. Inserted between the national finance capitalists and the oil enclave, this bureaucratic petty-bourgeoisie was important in defining the path and process of capitalist development. Its positive importance rested on its capacity to ignite the process of social change by contributing numerous activists, ideologues, and mass agitators.

The stratum, as a whole, had a negative importance also. First, because of its ideological affinity with the bourgeoisie and its concern with status, petty bourgeois radicalism finds its outer limits in programmatic statements calling for state enterprises and planning as bulwarks of national development. Second, the growth of this nonindustrial stratum has served to divert a substantial part of the nation's surplus value from productive activity toward the consumption of luxury imports, services, and so on.

Unlike the revolutionary transformations in Europe that eliminated certain classes, the oil-based transformation promoted one set of activities (industry) within a larger parasitical and wasteful setting. The result is that industrialization takes place at an enormous social and financial cost. Those at the top of the income distribution continue to skim off the benefits; those at the bottom are excluded. The emerging social structure is a patchwork quilt of continuing service, commercial, and bureaucratic strata sandwiched between the new finance groups.

The industrial home market was in part stimulated by overseas foreign investment and was not simply a creature of indigenous capital. In Venezuela there has been almost a complete absence of a tug of war between national and foreign capital within the industrial sector. The long-standing ties and associations dating back to the origins of the import-substitution strategy ensured peaceful collaboration. Clearly, industrialization through import substitution had a different meaning in Venezuela than in most of the rest of Latin America, where this strategy involved the stimulation and encouragement of national capital as well as the growth of an internal market and a reduction of consumer goods imports. Because this process began or was accentuated during the depression and World War II, little foreign capital was available to the process during its initial phase. On the other hand, Venezuela began its process of import-substitution industrialization in the early 1950s, during a period of worldwide expansion of U.S. capital. As a result, the measure did not encourage national capitalist development so much as it caused U.S. multinational corporations to jump tariff barriers and establish subsidiaries. The early involvement of U.S. enterprises within the Venezuelan market fostered multiple ties with local entrepreneurs and ensured that future nationalist movements would not be able to elicit the support of the local businessmen. This lesson has not been lost on the oil companies, which today remain associated with local capital in numerous capacities.

CLASS AND STATE

While the proposition that the state represents a ruling class is accepted, it must also be recognized that in a heterogeneous social formation such as Venezuela there are diverse factions within the bourgeoisie. At times, different factions have coalesced into a dominant power bloc, which has used the state's control over petroleum to further its own politico-economic power at the expense of its rivals within the Venezuelan ruling class; changes in hegemony were manifested in the rise to power of agricultural and then commercial and banking-industrial interests—first in alliance with English and later U.S. trading or corporate enterprises.

Moreover, ruling class political representation has been mediated by a diversity of sociopolitical forces reflecting regional and institutional interests. The military and the rising professional petty bourgeoisie each developed its own style of governing. While overall economic growth was determined largely by the changing demand for oil and the overall income distribution has remained the same, the type of growth has been influenced by the regime; the shift from Pérez Jiménez' military dictatorship to AD-governed democracy increased the profit-taking opportunities of national capital and the salary levels of the professional strata and allowed for the proliferation of a civil service bureaucracy through an AD-dominated patronage system.

Throughout Venezuelan history, the state has played a major, albeit changing, role and has been the key factor in the centralization and concentration of the capital accumulation process. As in other Western countries, the practice of state intervention on behalf of private capital is accompanied by the laissez-faire rhetoric of the capitalist ideologues. In actual practice, state intervention for business has been considered proper, while intervention on behalf of labor is interference with the self-regulating free market. With the exception of the trienio, all the post-Gómez government activities have complemented those of the private sector, not supplanted it, notwithstanding occasional protests from the business community. The proliferation of state agencies and autonomous corporations has not been the expression of encroaching socialism, but rather the development of a form of state capitalism in which the long-term, large-scale upstream investments in basic industry lay the basis for the smaller, high-return downstream manufacturing enterprises. Early industrialization in Venezuela has involved heavy state investments, but these basic industries, through administered prices, have served to subsidize private capital, as for example, in the growth of low-cost infrastructure facilities.

The factors central to the capitalist class are access to the state and government intervention on behalf of its interests. The access that the 1945 coup momentarily provided to a precocious petty bourgeoisie was assured to the national industrial bourgeoisie in 1958. The first AD regime attempted to balance off the interests of labor and capital, but the concessions to the former limited the willingness of the latter to cooperate. The second AD period witnessed an attempt to reconstruct the multiclass alliance, with the important difference that business and commercial interests were clearly hegemonic within the coalition. In the postwar interim (1948–57), the government actively promoted foreign capital in jointly expanding the infrastructure. Later, during the 1959–68 period, the governments increasingly intervened through import quotas, exchange controls, and financial aid to promote national industrial capital. From 1969 to the present, the emphasis has been on the encouragement of private sector activity through the creation of state and mixed enterprises, working toward a broader based and more diversified export sector.

THE NATIONALIZATION

The growth of state activities has paralleled the expansion of capital. The increased expenditures of the state in large part reflect the growing scope and reach of investors. It is within this context that one can understand the timing of the nationalization measures. The latter became feasible (and possibly necessary) for the state only when a fully developed bourgeoisie had emerged that could absorb the added funding, that is, channel the new financial resources into their own projects. To have nationalized earlier would have meant, as Betancourt himself admitted, a "leap into space." It would have required the AD to develop through the state a largely publicly owned industrial sector, preempting profitable economic "space" from the emerging bourgeoisie. This would have been incompatible with its own professed goals. Even the limited changes initiated by the first AD government were fraught with severe tensions due to the fears and hostilities of the landed-commercial and foreign oil interests, which were generated by the mobilizations of the urban and rural masses.

The collapse of nationalist-populist mobilization set the stage for the emergence of the latter-day entrepreneurial orientation of the Venezuelan governments. In the more recent period, the drive toward national industrialization from above has set forth a set of contradictions, the resolution of which will determine the direction in which the Venezuelan social formation will head in the near future. The general antagonism that will provide the revolutionary left with a multitude of specific issues around which to organize is the gap between Venezuela's potential for social development (equity, growth, welfare) due to the increased access of the national bourgeoisie to the nation's oil wealth and this class's inability to realize this opportunity because of its unwillingness to break the fetters inherent in the nation's "location" within the framework of world capitalist accumulation. Concretely, the most important conflict of interests in this regard that can be expected to arise will be between bourgeois regimes oriented to external markets, foreign capital, and U.S.-dominated international banks, on the one hand, and rural and urban masses demanding redistribution of property and income, increased living standards, and increased state investment-expenditures on social capital and services on the other.

A further set of subsidiary, although potentially acute, contradictions will be apparent: different factions of finance capital will contest the allocation of state investment funds, subsidies, mixed ownership, partnerships, import licenses, foreign exchange, and so on; capitalists will engage in political battles with technocrats within the state enterprises over their respective shares of the market, areas of private investment, conditions of association and ties with foreign corporations, and so on; and a general struggle will ensue between the petty bourgeoisie and the national bourgeoisie over the division of the national product between consumption and investment.

The success of the Venezuelan bourgeoisie in managing such contradictions depends on its continued control of the state. This will require the fashioning of a political program that combines sufficient expansion in government services to accommodate the petty bourgeoisie and the maintenance of a vast patronage network to contain discontent in low-income residential areas. Without some sort of vertical coalition, the bourgeoisie clearly will be unable to sustain its rule within the current parliamentary framework.

The principal threats to this project are from three directions: actions by the U.S. imperial state and indirect pressure by U.S. capital, which could lessen the autonomy of the Venezuelan state, possibly undermining its capacity to finance the projected expansion; the fragmentation and intensification of conflict between ruling class factions and the bureaucratic stratum over the allocation of state revenue, leading to rivalries in which the dominant capitalist parties cease to control the political agenda of the electoral system; and the emergence of a cohesive and class-conscious working class allied to the small peasantry and organized around a political party of the left or a bloc of such parties.

The first threat conceivably could undermine the stability of the present government; however, there are two aspects to this antagonism. On the one hand, while the U.S. government continues to express its opposition to OPEC and Venezuela's role in it. Washington has in no way taken any actions that would be likely to jeopardize directly the Pérez regime, let alone that might bring a popular leftist coalition to power. Nevertheless, U.S. capital clearly has increased its penetration of the Venezuelan social formation since the nationalization through the multinational corporations and the U.S. imperial state's influence in the international banks.

The growth of national industries in the context of an expanding foreign sector bodes ill for the future (see the Appendix). As oil revenues decline and as the foreign sector meshes with local industry, the possibility of large-scale foreign takeovers is ever-present. The current system of parallel growth is based on increasing oil revenues to finance national capital against the superior resources of multinational corporations. There are already signs that the Venezuelan state's capacity to serve as an independent national force is weakening.[1] The move toward foreign loans in 1976 and 1977 (see the Appendix) in order to finance large-scale property may be a harbinger of new dependent relations.

In all, the recent success of the U.S. bourgeoisie in tying the oil nationalization closely to its own project of capital accumulation presents the Venezuelan revolutionary left with ample evidence to rip away the mantle of nationalist legitimacy with which the current regime cloaks itself.

SUMMARY

The second potential threat to the state's stability is unlikely to lead to any conflicts of an insurrectionary sort, as political competition of this kind will probably be contained within the boundaries pertaining to the allocation of expenditures by the autonomous agencies and state enterprises and the degree and extent of state intervention. There have been some recent shake-ups in the Pérez government,[2] but despite large-scale mismanagement of funds and haphazard administration, the diversification of the economy should continue to produce new segments of the capitalist class. As has been the case in Mexico, for example, many may use their bureaucratic positions as springboards for their entrepreneurial ventures.

National dependence on, or exploitation by, foreign (largely U.S.) capital will continue to shape political issues and contribute to crises of accumulation within Venezuela. However, it is expected that the contradiction between national labor and capital will be the principal one in the future. What has been the effect of the nationalization on this conflict and which class will represent the motor force of Venezuelan social change in the period immediately ahead? Unlike many Third-World countries, the Venezuelan nationalization, rather than being based on mass mobilization, was tightly controlled and orchestrated from above; the process was low key and essentially tied to the initiatives and directives of the pro-industrialist leadership of the AD. In no important sense did the populace directly intervene in the process through its own organizations and set the terms or scope of the nationalization; much less do the popular classes have any say in how the state enterprises operate. Because the nationalization process did not awaken popular passions, the subsequent changes made in the sphere of production provided few opportunities for changes in the level and scope of popular influence over national politics.

Oil wealth, then, by expanding the opportunities for both national and foreign capital, has served as a substitute for an anti-imperialist movement. Also, the revolutionary leadership of the working class remains badly divided and until now has been unable to devise a program to build a coherent, sustained political force. Given the present weakness of the Venezuelan labor movement, it appears that the third potential threat to bourgeois political stability will not be realized in the immediate future. However, the very absence of any substantial working class pressure on the Venezuelan state has meant that the entrepreneurial formula of economic growth based on government incentives and subsidies to foreign and national capital has apparently led to increasing social inequalities (see the Appendix).

Clearly, nationalization is no panacea for the social ills that face Third-World countries, and there is no necessary connection between the state taking over private enterprise and the alleviation of un- and underemployment, inadequate health care and education, unequal wealth and income, exploitation, and so on. Given that these problems continue to exist in Venezuela today, the objective conditions for building a united mass movement are present. An understanding of the class nature of the nationalization may help strengthen the leadership of this movement. Because the benefits of the nationalization have been so one-sided, it will be harder for reformists to maintain that the state is a neutral institution capable of arbitrating class conflicts. This may make it easier to drive a wedge between the Ministry of Labor and the conservative labor leadership in the trade unions and the rank and file workers. Due to the manifestly weak role that Venezuelan finance capital played in bringing about the petroleum takeover, and in light of its continuing partnership with U.S. capital, there should be a greater degree of political clarity among the Venezuelan left concerning the policy of attempting to form tactical alliances against imperialism (though this has proven to be a correct policy in other contexts) with the "progressive" national bourgeoisie.

Finally, since both the iron and petroleum nationalizations have pushed Venezuela much farther in the direction of state capitalism, it will be easier in the future to combat economism among oil and steel workers, whose direct antagonist will be the bourgeois state itself—thereby making every economic issue a political one as well.

THE UNITED STATES, VENEZUELA, AND THE INTERNATIONAL POLITICS OF OIL

The dependence of the U.S. economy on oil imports to meet internal consumption requirements increased from 36 to 42 percent of total U.S. oil needs between 1973 and 1976, and it is estimated that the figure will reach 51 percent in 1977.[3] It was in this context that U.S. government hostility to OPEC continued throughout 1976 and that there was an intensified attempt to pressure and divide the OPEC membership in order to contain oil price increases. While conceding that "[its] margin of decision [was] relatively limited,"[4] the Ford administration focused its maximum resources on a strategy to disintegrate the organizational focal point of Third-World efforts to equalize the terms of international exchange on the basis of its oil wealth.

In late November of that year, then Secretary of State Henry Kissinger sent a confidential telegram to the European delegation to the Conference on International Economic Cooperation (CIEC) meeting in Paris—more popularly known as the North-South dialogue of industrialized, developing, and oil-producing countries—which was aimed at elaborating the contours of the so-called new international economic policy. Kissinger expressed adamant opposition to any proposed concessions by the industrialized countries in exchange for "moderate" oil price increases at the December meeting of OPEC ministers in Qatar, stating that "(t)his would reverse the linkage we would be seeking, and would strengthen OPEC ties to other LDC's [less developed countries]."[5] He was quite explicit in the need to weaken Third-World unity on this issue:

> We are convinced that there is no negotiable CIEC package which the industrialized countries could accept and which would also represent sufficient inducement to OPEC to refrain from a substantial oil price increase over several years, given the lack of leverage by consumers over oil prices. While the oil price decision can affect our ability to take actions responsive to LDC proposals in the North-South dialogue, we must discuss these issues on their own merits in CIEC. The linking of CIEC and OPEC could undermine this effort, making decisions in OPEC depend on decisions in CIEC, rather than the reverse. *We have been relatively successful in CIEC in intensifying LDC restraints on OPEC* (emphasis added).[6]

It is within this context of international political maneuvering that the U.S. government's appraisal of the class nature of the Venezuelan government that has carried out the oil nationalization has been proven correct and that is now beginning to pay off. Venezuela has continued to provide the United States with unfettered access to its oil resources—allowing the latter to absorb approximately 70 percent of the former's oil exports.[7] The Venezuelan government has continued to give its qualified support to a strategy linking future oil prices increases "to the degree of imbalance existing in the price of oil and the price of manufactured goods imported by the developing countries," but has also opposed "radical" changes proposed by OPEC members as "impractical and unfeasible."[8]

Furthermore, Venezuela's conciliatory and "moderate" role within OPEC was most recently reflected in its advocacy of an intermediate 10 percent oil price increase at the December 1976 meeting of OPEC ministers. Perhaps Pérez himself has expressed the position of his regime as clearly as anyone when he stated that oil is "an instrument of negotiation and not of confrontation."[9] In the eyes of U.S. capital, in this period of world capitalist crisis and national liberation movements, Venezuela must be coming to appear as though it were a veritable port in the storm.

NOTES

1. See "Venezuela: Tax Worries," *Latin America,* December 24, 1976. For example, the resignation in December 1976 under pressure from the Venezuelan right (AD, COPEI, Fedecámaras, the finance minister) of one of the more aggressive nationalist government officials, ex-Minister of Planning Gumersindo Rodríguez.

2. "Venezuela: New Team, Old Party," *Latin America,* January 14, 1977.

3. See William D. Smith, "Oil Experts Say an Embargo Now Would Hurt United States More Than in '73." New York *Times,* October 18, 1975, pp. 1, 48; Edward Cowan, "... And

Still U.S. Energy Alternatives Are Weak," New York *Times,* December 5, 1976, pp. F1, F7; "Record U.S. Oil Imports Seen This Year, 51% of Total Used," Washington *Post,* February 20, 1977, p. A20.

4. Henry Kissinger, quoted in Ann Crittenden, "Kissinger Sees U.S. in Limited Ability to Stem Oil Prices," New York *Times,* November 19, 1976, pp. D1, D7.

5. Text of Kissinger telegram published in its entirety in Peter Rodgers et al., "Double Pressure for Oil Price Rise," London *Sunday Times,* December 12, 1976, p. 53.

6. Ibid.

7. See U.S. Department of Commerce, Domestic and International Business Administration, Bureau of International Commerce, *Venezuela: A Survey of U.S. Business Opportunities,* Country Marketing Sectoral Survey, June 1976, p. 1.

8. Carlos Andres Pérez, quoted in Peter Grose, "Venezuelan Leader Ties Oil Rise to Accord at North-South Talks," New York *Times,* November 17, 1976, pp. D1, D11.; see also Carlos Andres Pérez, "Altering the North-South 'Collision Course'," New York *Times,* December 15, 1976, p. A25.

9. Quoted in Gregory F. Treverton, "Venezuela and Its World Role: A Conversation with President Carlos Andres Pérez," *WorldView,* no. 9, September 1976, pp. 39–42.; see also "Pérez Opposes Leap in Price of Oil," Washington *Post,* November 20, 1976, p. A5.

APPENDIX: RECENT TRENDS

NATIONALISTS WITH MONEY

The increased national wealth associated with the nationalization has meant economic growth of a particular type and increased penetration of foreign (especially U.S.) capital. According to the U.S. Department of Commerce, since 1973, Venezuela has experienced "only modest real growth,"[1] and many announced development programs are reported to have achieved disappointing results:

> Venezuelans have begun to ask where all the money has gone when development projects are barely off the ground. The growing gap between expectation and achievement has caused widespread disillusion that is compounded by reported instances of graft and corruption within the administration. The country is also set back somewhat by its huge propensity to import. Prompted by its oil riches, Venezuela has now developed a voracious appetite for the goods that these funds can buy. The industrial base, despite some efforts at decentralization away from oil, has not advanced as strongly as it could have if more resolve had been applied.[2]

In 1975, foreign imports rose by approximately 35 percent, while exports increased by just over 1 percent. Given present trends, this imbalance is likely to continue in the foreseeable future. Imports climbed by an estimated 15 percent in 1976 and are expected to rise by 12 percent in 1977—both figures in excess of export increases for this two-year period.[3] The failure to reduce the level of purchases of luxury and nonessential imports by the Venezuelan upper income stratum has raised the specter of balance of payments deficit for the first time in a number of years.

In addition, stagnation in the agricultural sector, despite large amounts of financial assistance from the state, has necessitated the importation of basic foodstuffs. In 1975, for example, the government budgeted $1.5 billion for a variety of agricultural projects which did not include public works expenditures that directly benefited commerical farmers and cattle ranchers. Nevertheless, Venezuela was forced to import over $1 billion worth of food supplies, which accounted for almost 20 percent of total imports in 1975.[4]

Under concessions signed with the government, the foreign oil companies continue to do very well in Venezuela, retaining control over the distribution, marketing, and transportation operations in the postnationalization period. The profits flowing to these corporations have been estimated to be in excess of those prior to nationalization. During 1976, for example, Gulf Oil Corporation reported that worldwide refining, marketing, and transportation operations amounted to approximately 20 percent of its total profits for the year, as compared with break-even operations during 1975.[5]

In general, both U.S. trade with and direct investment in Venezuela are expanding rapidly. The value of U.S. exports to Venezuela increased from $2.24 billion (47 percent of total Venezuelan imports) in 1975 to $2.63 billion in 1976, and U.S. policy making and business officials are preparing for an even greater increase in 1977:

> Venezuela is assignd a high priority in the United States export expansion efforts under the "sell in Venezuela" campaign, which involves greatly expanded market research, a schedule of almost twenty trade promotion events, and intensified pursuit of major project opportunities. The government's intention to build up productive infrastructure with public investments on the order of $20 million through 1980 implies substantial opportunities for United States consultants, engineers, and industries which can supply process, licensing and capital goods, in many cases on a turn-key basis, as well as heavy construction contractors.[6]

The expanding requirements for technology, equipment, and management expertise under the national development plan has created a market "for a broad spectrum of imports from the U.S. [which] will continue strong and the predicted high level of foreign exchange availability over the foreseeable future will ensure adequate means of payment."[7]

The expansion of new capital investment opportunities in downstream growth sectors of the economy, especially manufacturing (steel, petrochemicals, and aluminum), transportation, power, and even petroleum, has pushed total U.S. direct investment in Venezuela to over $2 billion.[8] Within the manufacturing sector, capital expenditures by majority-owned foreign affiliates of U.S. companies are expected to increase by almost 100 percent between 1976 and 1977, from $118 million to $226 million.[9] Among the beneficiaries of this expansion of manufacturing, concentrated in foodstuffs and chemical products, are U.S. oil companies; it is suggestive in this regard that Gulf Oil Corporation increased its overall chemical earnings in 1976 18 percent over the previous year's level.[10]

INTERNATIONAL DEBT PEONAGE

The recent decline in the amount of economic surplus available for development projects has resulted in a substantial increase in external financing of particular investment projects and the curtailment or elimination of funds for other projects initially promoted by the Pérez government. In September 1976, the Venezuelan government borrowed $1 billion through an international banking consortium led by U.S. Morgan Guaranty Trust and Citcorp in order to refinance the short-term debts contracted by the autonomous public agencies. In January 1977, it moved to obtain large-scale foreign financing for the fifth national development plan in the form of a syndicated eurodollar loan accepted by an 18-bank consortium, again headed by two major U.S. banking corporations, Morgan Guaranty Trust and Chase Manhattan Bank. This loan amounted to $1.2 billion and was augmented by a further $100 million eurobond issue managed by a banking group under the direction of the First Bank of Boston (Europe).[11]

Although Venezuela has a current foreign reserves cushion of approximately $8 billion to cover this external borrowing, any decision by the Pérez government to maintain this trend could quite conceivably lead to an uncontrollable debt service burden in the event that "the industrialization programe fails to produce sufficient returns."[12] It is estimated that the debt service burden will reach 15 percent of the total budget by the end of 1977 as compared with 6 percent in 1975, and that the total direct and indirect national debt will more than double over the next four years. By that time, the total national debt will approximate 60 percent of the country's present international reserves and assets of the Venezuelan Investment Fund combined.[13]

THE LIMITS OF TRICKLE-DOWN GROWTH

Limited growth in a context of overall relative economic stagnation has been paralleled by government policies that have done little to alter decisively the preexisting class and regional economic inequalities in Venezuelan society. The selective and short-term redistributive policies of Pérez' populist period resulted in marginal socioeconomic gains for the working class, but has since been followed by no significant working class legislation. Incremental or relative improvements in the areas of social welfare and housing have not significantly improved the conditions of life of the urban and rural proletariat, and their incapacity to make demands at the national level vis-a-vis other classes in society suggests little likelihood that this situation will change in the near future. A recent World Bank report discussed the development of health facilities under the Pérez government:

> Curative health programs are quite inadequate at present, and current plans promise no significant improvement. Few aspects of Venezuelan life seem to evoke such widespread worry and complaint as health care, and much time seems to be lost due to ill health. Private medical services are expensive, and private hospital facilities are scarce. Public hospitals and curative facilities

are not only quantitatively insufficient, but the services often appear poorly designed and administered. Dissatisfactions are most acute in relation to the Social Security medical system.[14]

Although the proportion of houses with running water and sewerage facilities has gradually increased over time, "available data suggests that at least one-quarter of all houses are crudely and inadequately constructed, even by modest standards. In addition, most houses are crowded. Over one-half of all family housing units in 1974 sheltered more than twice as many people as the number of bedrooms; even this ratio has been increasing slowly over time."[15] On a regional comparison, the quality of, and facilities available to, rural housing is considerably inferior to that of urban dwellings.[16]

The cost of living and monetary supply statistics for 1976 do suggest some overall success by the government in controlling the persistent problem of inflation. The growth in the money supply declined from 47.4 percent in 1975 to 24.5 percent in January to October 1976, while the rise of 7.7 percent in consumer prices between January and November 1976 (as measured by the Caracas cost of living index) compared favorably with a 10.3 percent increase for the similar period in 1975. However, these gross figures are deceptive in that they underestimate the growth in the cost of basic consumer necessities over time (see Table 5).

TABLE 5

Summary Trends in Venezuelan Prices
(annual average percent change)

	1972	1973	1974	1975	1976
Cost of living in Caracas	2.9	4.1	8.3	10.3	7.7
Food, beverages, and tobacco	4.9	7.6	12.7	14.3	—
Clothing and shoes	2.9	7.7	20.2	17.0	—
Housing	1.5	2.2	4.5	5.0	—
Other	2.4	2.4	6.2	10.1	—

Source: Latin America Economic Report, "OPEC's difficulties cloud Venezuela's prospects," January 14, 1977; International Monetary Fund, Venezuela—Recent Economic Developments, Confidential Report No. SM/76/89, May 12, 1976, p. 15.

Despite a slight decrease between 1974 and 1975, price increases for food and clothing remain significantly above the average cost of living rise and well in excess of cost increases for these items during the pre-Pérez period. The impact of these particular increases is likely to have had a disproportionate incidence in the lives of working class people and although comparative figures are not available for 1976, it is doubtful if this trend was reversed.

SUMMARY

BEHIND THE NATIONALIST FACADE[17]

The promise of the oil nationalization is falling far short of the expectations of many Venezuelans. Thus far we have emphasized the class nature of the distribution of the newly-acquired sources of oil revenue and the reinsertion of foreign capital into the growing Venezuelan home market. There is, however, an additional factor to explain this shortfall in anticipated benefits, which also accounts for the relative equanimity displayed by the officials of the expropriated foreign petroleum corporations and of the U.S. state: the agreed-upon compensation package greatly favors these corporations and places considerable limits on the total amount of net oil revenue which will accrue to the Venezuelan state for the purposes of discretionary government spending over the next few years.[18]

In effect the total compensation includes both the indemnity payment for the expropriation plus the fees for the "technical assistance" the oil corporations have contracted to provide. While the former is substantial in itself, the "service fees" are roughly equivalent to the revenue which these corporations could have expected from their investments in the period between the nationalization (1976) and the date for reversion of the concessions and installations (1983).

By any standard the Bs.4,300,000,000 indemnification was extremely generous. From 1965 to 1975 the total average net investment for the oil corporations was maintained at roughly Bs.7,000,000,000, while aftertax profits in the same period equalled almost Bs.60,000,000,000.[19] In other words, in the ten years leading up to nationalization the petroleum corporations had recouped their original investment (excluding depreciation) eight and one-half times! Furthermore, since these companies always operated on the basis of short-term profit maximization, they had passed on the "social costs" of oil extraction to the Venezuelans. Therefore, it could be logically argued that an amount equal to the accumulated loss of natural gas and the ecological damage to the Maracaibo basin from oil spills should have been deducted from the indemnification.[20]

While this form of compensation has been public knowledge, it appears as though there has been a conscious attempt to hide the terms of the "technical assistance" contracts in an effort to disguise the onerous nature of the overall compensation.[21] In order to appreciate the exact nature of these agreements it is necessary to understand how they are drawn up and carried out. PETROVEN functions as a state-owned "holding company" for the various state oil enterprises (LAGOVEN, LLANOVEN, MARAVEN, TALOVEN, MENEVEN, and others), which actually operate the concessions and installations that were formerly maintained by the foreign oil corporations. The president of PETROVEN, Rafael Alfonzo Ravard, signs each of the agreements along with a representative of the particular oil corporation involved

(Shell, Mobiloil, Exxon, etc.). The actual execution of the contract is then carried out by two parties: the state oil enterprise charged by Ravard with the responsibility and the "service company" created by the oil corporation (Mobiloil's MODECA, for instance). Although the contracts vary somewhat according to the size and nationality of the oil corporation (North American or European) and the production level of the state oil enterprise, in general the terms are very similar.

What then do the contracts actually provide the Venezuelan state? Based on the terms of the LLANOVEN-MODECA agreement, *Proceso Politico* concludes:

> The contracts establish that assistance in production operations is to be limited to the area of the previous concessions and any assistance in the refining processes is to be limited to the refining capacity extant at the moment of the nationalization. The use of the same technology in areas outside the previously existing concessions or the amplification of plant refining capacity must be the subject of *separate agreements*. The contracts explicitly exclude the use of the technology necessary for the exploration, production and refining of heavy crude oils, which constitute the nation's most important reserves. The use of this technology must also be purchased via separate agreements.[22] (translation ours)

Contrary then to the rationale used to justify these agreements, *the contracts do not provide for the creation or transfer of new technology to the state oil enterprises*. Rather, they merely grant permission to apply techniques which are currently known in Venezuela for the duration of the contracts. Furthermore, in many instances the diffusion of technology has been, and will continue to be under the present agreements, from Venezuelan technicians to other regions exploited by the oil corporations.[23]

All the contracts stipulate ceilings on the number of man-hours per year of advising that are to be provided, with separate contracts required for any "overtime." The travel expenses of all the service company personnel are to be met by PETROVEN; even more importantly, all the operating expenses which MODECA, for example, incurs in Venezuela while conducting business with LAGOVEN and LLANOVEN are to be covered by these state enterprises, while TALOVEN—due to its smaller level of oil production—pays a fixed rate of $45,000 per month for such costs.[24]

The service companies are absolutely free from any legal responsibility for the consequences of any "bad" advice which they may provide. The state oil enterprises are, of course, free to ignore the consultation, but should they discover that they have been following policies oriented to the business interests of the petroleum corporations rather than their own, they will have no way to recover past "losses." There are several ways in which this could easily happen, given that the service companies can be expected to act in the interests of their parent corporations. For instance, they are likely to influence the direction of Venezuelan petroleum-related industrialization in terms of both the kinds of research performed and the types and sources of capital goods imported. This influence will increase the danger of overpriced imports and also that of the "dumping" of obsolete equipment by the parent corporations. Furthermore, the efforts of oil conservationists may well be undermined by the service companies, who will be concerned with maintaining abundant petroleum exports at low prices.

The rules governing the conduct of business relations between the service companies and the state oil enterprises serve to reduce the sovereignty of the Venezuelan state and to fragment its negotiating capacity *vis a vis* the oil corporations. Neither PETROVEN nor any particular state enterprise can legally pass on information regarding any business transacted with a given service company, while it will be impossible to prevent the service companies from collaborating among themselves. Also, there exists the mutual right to inspect the books of one's contracting "partner." This reciprocity is not equitable, however, considering that while PETROVEN is merely a holding company, the parent oil firms can withold information on their overall worldwide operations and their collusion behind the "shield" of the service companies.

In the formula devised to calculate the *total payment per barrel* which is to be paid to the service companies for their "technical assistance," the tax rate, the exchange rate, and the volume of 'production' of crude and refined oil can, in theory, all vary. The *net profit per barrel* received by these companies is, however, a constant which is fixed by the terms of the contract. The total payment is treated as an operating expense and is met by the state oil enterprises. Since the net profit per barrel is fixed and the total payment per barrel varies inversely with the corporate tax rate, if the Venezuelan government were to increase the tax rate paid by the service companies (currently 50 percent), the state oil enterprises would have to increase the amount of their total payment to the service companies. The incidence of the tax would then fall on the state enterprises and the Venezuelan government will have in effect renounced its power to determine taxation levels.[25]

Although the contracts run for four years, the net profit per barrel guarantee is renegotiated annually to assure that it increases apace with the rate of "world inflation." This yearly adjustment is made on the basis of the "Imported Products Price Index" calculated by the Central Bank of Venezuela. This can only be regarded as further evidence that the "technical assistance" agreements constitute a covert form of compensation for the expropriated oil corporations. The renegotiated profit rate does not indicate any increased level of service to the Venezuelan state; neither can it be considered a reflection of increased costs incurred by the service companies, since all personnel costs and travel expenses are paid by PETROVEN and the cost of imported capital equipment is met by the state oil enterprises.

In 1976, based on the government's own (albeit varying) calculations, the "technical assistance" contracts cost the Venezuelan state approximately Bs.700,000,000 after taxes. Leaving aside the questions of increasing production levels, special separate contracts, and renegotiated profit rates in the course of four years, the Pérez government will have obligated the Venezuelan taxpayers to provide the foreign petroleum corporations with another two-thirds the amount of the initial indemnification in return for precious few benefits.[26] The foreign oil firms, on the other hand, are guaranteed higher net profits than they would have enjoyed had they retained their installations and concessions—at no risk and with the Venezuelan government providing the investment capital—for merely helping maintain their previous operations at the same level of productive efficiency.

NOTES

1. U.S., Department of Commerce, Bureau of International Commerce, *Foreign Economic Trends and Their Implications for the United States: Venezuela*, 76–146, December 1976, p. 4.

2. "Focus on Venezuela: The Rainbow and Pot of Gold Are Beginning to Fade," *Business Latin America*, January 28, 1976, p. 27.

3. "OPEC's Difficulties Cloud Venezuela's Prospects," *Latin America Economic Report*, January 14, 1977, p. 7.

4. Joseph Mann, "Venezuela's Agricultural Ills," New York *Times*, January 30, 1977, p. 42 (international economic section).

5. Clare M. Reckert, "Gulf Oil Profits Up 28.2% in 4th Period," New York *Times*, January 28, 1977, p. D9.

6. On the value of U.S. exports to Venezuela, see U.S., Department of Commerce, Bureau of the Census, *Highlights of U.S. Export and Import Trade,* FT990/December 1975, Table E3, p. 40, and figures supplied by the U.S. Department of Commerce for inclusion in the 1976 edition of that report; on Venezuela as a high priority market, see, U.S., Department of Commerce, Bureau of International Commerce, op. cit., p. 4.

7. Ibid.

8. "U.S. Direct Investment in LA and Rates of Return Doing Better Than Elsewhere," *Business Latin America*, October 13, 1976, p. 322.

9. U.S., Department of Commerce, *Survey of Current Business,* September 1976, pp. 26–27; "Venezuelan Decree Softens Foreign Investment Rules," *Latin America Economic Report*, February 25, 1977, p. 29.

10. Reckert, op. cit., p. D9.

11. "Venezuela Ready to Launch Foreign Borrowing Drive," *Latin America Economic Report*, January 7, 1977, p. 2; "Venezuela Secures Better Terms for Eurodollar Loan," *Latin America Economic Report*, February 4, 1977, p. 17.

12. "Venezuela Ready to Launch Foreign Borrowing Drive," op. cit., p. 2.

13. Ibid.

14. World Bank, Latin America and Caribbean Regional Office, *Current Economic Position and Prospects of Venezuela*, Vol. 1, The Main Report, Confidential Report no. 1268-VE, February 1977, p. 69.

15. Ibid., p. 70.

16. Ibid.

17. What follows is largely a partial summary of two articles from a recent special issue of *Proceso Politico.* In spite of the secrecy surrounding the terms of the nationalization agreement, the editors of this journal were able to obtain copies of the actual "technical assistance" contracts, thereby getting access to information which has been unavailable even to most officials of the current AD government. See "La OPEP, la nacionalización del petróleo y el imperialismo hoy," pp. 2-54 and "El secreto de los contratos petroleros de asistencia técnica," pp. 55-85 in *Proceso Politico,* 4-5 (enero-abril, 1977).

18. Previewed investment levels anticipated for the petroleum industry in the Fifth National Plan (1976-80) amount to Bs.4,704,000,000 per year. If the 1976 level of state revenue from the sale of oil remains at Bs.3,160,000,000, this implies a yearly deficit of Bs.1,400,000,000. Ibid., p. 50.

19. Ibid., p. 39.

20. Ibid., p. 41.

21. We infer this from the one-sidedness of the contracts and the fact that President Pérez has misinformed the public (speech to the Cabimas Civic Center in the state of Zulia, January 1, 1976). to the effect that the contracts last two rather than four years (1976-80). Ibid., p. 59.

22. Ibid., p. 62.

23. For example, the offshore drilling platforms currently in use in the North Sea of Scotland were first developed with the assistance of Venezuelan engineers working in the Maracaibo basin. Ibid., p. 67.

24. Ibid., pp. 78-80.

25. Ibid., pp. 74-75.

26. Ibid., p. 77.

INDEX

AD, 8, 19, 36–37, 41, 79–82, 84–85, 87–89, 91, 102, 155, (*see also,* Betancourt); Pérez Jiménez period, 19; *trienio* period, 8–9, 12, 15, 19, 24–25, 31, 154; 1957–63 period, 22–25, 35, 39, 154; 1964–68 period, 25 (*see also* Leoni); 1969–73 period, 37, 154, 164; 1974–77 period, 39, 65–66, 68–69, 157 (*see also,* Andrés Pérez)

agrarian reform, 12, 19, 24, 32, 35, 38, 65, 91

Agricultural Investment Fund, 65

"Allende's Chile", xv, 100, 102, 124, 127, 133–34, 138, 142, 144

Andean Pact: Andean Development Corporation, 75, 141; regulation of foreign capital, 28, 69, 105–06, 112, 124, 133; Venezuelan entry, 34, 38, 40, 125

Andrés Pérez, Carlos, 1, 38, 40, 62, 64–70, 72–74, 83–90, 97, 99, 101–03, 105–06, 112, 114–20, 122–35, 138, 141–44, 156–57, 159, 163; electoral coalition, 66, 84; nationalist populist period, 62, 66, 101, 105, 130, 163; state enterprise domain clarified, 50, 60–62, 68–69, 74–75

Argentina, 116

autonomous agencies, 35–38, 64–68, 70, 72, 154, 157 (*see also,* specific names)

banking system, 19, 21–23, 62, 66, 68–69; Agricultural Development Bank, 18, 36, 66; BCV, 14, 37, 68; BIV, 14, 21, 36, 37 (*see also* credit); foreign capital penetration of, 2–3, 9, 112

Betancourt, Rómulo, 14–15, 25, 31, 79, 81, 82, 85, 155

Boulton finance group, 4, 10, 42

bourgeois revolutions, 151–52

Brazil, 2, 62, 70, 74, 76, 85, 108, 116, 129

CADAFE, 20, 23, 35
Caldera, Rafael, 28, 37–39, 41, 122
Caribbean, 75–76, 91, 108, 134
Castro, Cipriano, 3
Central America, 42, 75–76, 91, 108, 134, 141; Central American Bank for Economic Integration, 75, 141
China, 86

class transformation, 151–53
CODESA, 36
Colombia, 112
commercial agriculture, 4, 5, 6, 8–9, 14, 16–18, 20, 26, 32, 38, 65, 81, 152; exports, 2, 3, 5–7; food imports, 16; foreign capital penetration, 17, 62–63
commercial bourgeoisie, 2, 7, 34, 152; comprador element, 42, 52, 152
Conference on International Economic Cooperation (CIEC), 159
COPEI, 15, 19, 23, 24, 26, 36, 37, 41, 66, 81, 84, 87, 107 (*see also* Caldera)
CORDIPLAN, 23, 36, 37, 60, 66–68, 72
CORPOINDUSTRIA, 65
Council of the Americas, 112
credit: commercial, 3–4, 9, 18–19, 32; consumer, 27, 28, 38; foreign debt, 3, 8, 22, 24, 156, 163; government provision of, 8–10, 13–14, 18, 33, 36–38, 41–42, 64–65, 74–75, 81, 114 (*see also,* petroleum nationalization; Venezuelan State, petroleum revenue/surplus); public debt, 4, 8, 36
Creole Foundation, 29
Creole Investment Corporation, 28
Creole Petroleum Corporation, 28–29
CTV, 12, 19, 24, 36, 41; Vargas, José, 80
Cuba, 25, 76, 86, 138
CUTV, 36
CVF, 14, 21, 23, 36, 37, 42, 66
CVP, 13, 20, 24, 25, 30, 35, 39

DIVIDENDO, 29
dual society, 6, 38

Egypt, 59–60
exploitation, rate of, 33, 68, 74–75, 84
exports, diversification of, 27, 38, 40, 44, 72, 75, 91, 155

FCV, 12, 25, 37
Fedecámaras, 11, 14, 22–23, 29, 34, 36–37, 40–41, 68–72
Fedepetrol, 41
finance capital, 8–10, 17, 29, 34, 37–38, 42, 72, 152, 155–56, 158; groups, 4, 10, 17–

19, 23, 43, 53–54, 57, 91, 151–53 (*see also,* specific names)
FIV, 35–36, 42, 64, 143, 163
Ford, Gerald, 102, 129, 158

Guzmán Blanco, António, 4

home market, 1, 3–5, 9, 16, 27, 31, 38–39, 71–73, 75–76, 132, 151, 153

IMF (International Monetary Fund), 25, 75, 125, 141
imperial state, xvii–xxi; Fanon on, xviii; Bukharin on, xvii–xviii; Poulantzas, xix (*see also,* U.S. imperial state)
import substitution, 3, 9, 16, 20, 23, 27, 38, 40–41, 68, 153, 154
income distribution, 5, 32–33, 154, 157, 163, 164
Indonesia, 119
Industrial Development Fund, 14
industrialization, Venezuelan: artisan, 5–7, 11, 18; foreign capital penetration, 9, 16–17, 20, 22, 27–31, 62, 69, 71–73, 76, 105–08, 126, 152–53, 156; manufacturing and mining, 5, 9, 11, 14, 17–19, 26–27, 37, 62, 70–71, 73, 152–53, 162
Inter-American Development Bank, 75, 132, 141, 142–44

Kissinger, Henry, 95–97, 101–03, 128–29, 158

LAFTA, 34, 40
landed oligarchy (latifundistas), 3–4; formation 8, 17–18, 32; political orientation, 15
Leoni, Raúl, 37, 39, 85
López Contreras, Eleazar, 13, 15

MAS, 66, 85, 87–91
Medina Angarita, Isaías, 10, 12–13, 15, 20
Mendoza, Eugenio finance group, 11, 14, 18, 22–23, 37, 42; Mendoza Foundation, 29
MEP, 25, 41, 87–88
Mexico, 59, 62, 116, 129, 157
military dictatorships, 2, 5, 8, 12, 15, 19, 43, 85; Patriotic Military Union, 12; provisional government, [1945–46, 12–13], [1958, 21–24, 35, 37]
MIR, 25, 35, 87–91

Mito-Juan, 30
Montana, finance group, 42, 70
Mossadegh, Mohammed, 77

National Security Council, 95, 100
NATO, 96
Nigeria, 119, 130

OPEC, 25, 39, 71, 77, 91, 97, 102, 104, 112, 125–26, 128–32, 135–36, 138–39, 145; Arab oil embargo, 39, 64, 96, 101, 139–40, 158–59

Pérez Alfonzo, Juan Pablo, 13, 25, 41
Pérez Jiménez, Marcos, 2, 15–22, 24–26, 30, 35, 38 (*see also,* military dictatorships)
Peru, 54, 70, 74, 85, 142
petroleum companies: investment, 5, 6, 108, 113, 115–16, 121–22; policies, 7, 13–15, 20, 39, 115–16, 118–19; relations with oil sector, postnationalization, 64, 70–71, 116–21, 159, 161–62
petroleum enclave, 5–6, 10, 13–14, 28–30, 106, 152
petroleum nationalization: state surplus acquisition/disposition [forms of, 58–59], [theory of, 50, 52–57]; Venezuelan regulations on foreign capital, 61, 69, 71, 72, 74–75, 107, 112, 159
PETROVEN, 42, 70, 72
petty bourgeoisie: formation, 7, 9, 11, 15, 31–32, 63–64, 86, 152; political orientation, 12, 15, 24, 31, 41, 43, 63, 66, 86, 152, 156
political parties (*see,* specific party acronyms)
political patronage, 15, 36, 59, 66, 79, 82, 88, 90, 156
political repression, 12, 19, 35–36, 74–75, 84–86, 89–90
price inflation, 67, 68, 83, 86, 106, 112, 145, 164
progressive national bourgeoisie, 2, 4, 10–11, 31, 38, 41–44, 52, 67, 69–71, 73, 78, 82, 158
proletariat: agricultural, 3, 5, 17, 32, 38, 57 [political orientation, 37, 86, 90, 156–57]; industrial, 3, 5, 7, 9, 11, 18–19 [political orientation, 11–12, 19, 24, 34, 36–37, 41, 66, 87, 90, 91, 152, 156–57]
Pro-Venezuela, 23, 34, 41

SELA, 76, 138
service sector, 5, 7, 10, 26–27, 32, 34, 78, 90, 92
Simon, William, 102
state capitalism: defined, 58; Venezuelan class and market constraints, 61
state, petroleum revenue/surplus: acquisition, 5, 7–9, 13–15, 19–20, 22, 25, 29–30, 39, 151 [postnationalization, 86, 114]; disposition, 9–10, 13–14, 17, 20–23, 27, 29–31, 33, 35–38, 41, 44, 72, 78–82, 154, 156 [postnationalization, 62, 65–68, 92, 143 (*see also*, relevant state agencies)]
state, relative autonomy of, 5, 22, 43; extent of state party penetration, 53–54, 74

Tinoco: Tinoco Convention, 8; Pedro, Sr., 15; Pedro, Jr., 37, 42
trade unions, 12, 36, 41, 87–89, 152, 158 (*see also,* proletariat)
Turkey, 59–60

unemployment/underemployment, 6–7, 11, 19, 25, 27, 32–34, 57, 83, 85–87, 90
urbanization, 5–7, 9, 18, 24, 31
URD, 15, 19, 23, 24, 41
U.S. imperial state: agencies, 103–04; policy formation, 95; policy goals, 95–97, 158–59; and Third-World nationalizations, 100, 156; and U.S. multinational corporations, 97–98, 121–23, 131
U.S.-Venezuelan trade, 38, 107–08, 110, 112–14, 126–30, 32–33, 161–62; food imports (*see,* commercial agriculture); U.S. oil dependence, 96, 101–03, 124, 127–28, 158; U.S. oil import quotas, 39; Venezuelan military sales credits, 132, 136–37; Venezuelan tariff preferences (U.S. Trade Act, 1974), 102–03, 113, 128–30, 135, 139

Venezuelan Association of Executives, 29, 34, 41
Venezuelan Communist Party (PCV), 24, 35
Venezuelan Development Fund, 37
Venezuelan Petrochemical Institute (IVP), 21, 35, 37
Venezuelan petrodollar loans, 75–76, 110, 125 201, 214, 220
Vicente Gómez, Juan, 3, 5–8, 10, 12, 15, 16, 154
Vollmer finance group, 3, 10, 18, 22, 42

Williams, Eric, 76
World Bank (IBRD), 25, 75, 101, 125, 132, 141–45, 163

Zuloaga finance group, 3, 18, 37, 42

ABOUT THE AUTHORS

JAMES F. PETRAS has written extensively about Latin American political, economic, and social issues, and about the influence of the United States on Latin American affairs. Among his recent books are *U.S. Imperialism and the Overthrow of Allende* (Monthly Review Press, 1975), *Latin America: Dependence or Revolution* (Wiley, 1973), *Peasants in Revolt* (University of Texas Press, 1973), and *Cultivating Revolution: The United States and Agrarian Reform in Latin America* (Random House, 1971). Dr. Petras is professor of sociology, State University of New York, Binghamton.

MORRIS MORLEY is a Ph.D. candidate in the sociology department at SUNY Binghamton.

STEVEN SMITH is a graduate student at SUNY Binghamton.

RELATED TITLES
Published by
Praeger Special Studies

OIL IN THE ECONOMIC DEVELOPMENT OF VENEZUELA
      Jorge Salazar-Carrillo

ECONOMIC GROWTH AND EMPLOYMENT PROBLEMS IN VENEZUELA: An Analysis of an Oil-Based Economy
      Mostafa F. Hassan

*VENEZUELA: The Democratic Experience
      edited by John D. Martz
      and David J. Myers

ECONOMIC NATIONALISM IN LATIN AMERICA: The Quest for Economic Independence
      Shoshana B. Tancer

EXPROPRIATION OF U.S. PROPERTY IN SOUTH AMERICA: Nationalization of Oil and Copper Companies in Peru, Bolivia, and Chile
      George M. Ingram

EXPROPRIATION OF U.S. INVESTMENTS IN CUBA, MEXICO, AND CHILE
      Eric N. Baklanoff

ALLENDE'S CHILE
      edited by Philip O'Brien

---

*Also available in paperback as a PSS Student Edition.